T0340180

Bio-Inspired Algorithms
FOR ENGINEERING

BIO-INSPIRED ALGORITHMS
FOR ENGINEERING

ALMA Y. ALANIS
NANCY ARANA-DANIEL
CARLOS LÓPEZ-FRANCO

University of Guadalajara
University Center of Exact Sciences and Engineering
Department of Computer Sciences
Intelligent Systems Research Group

Butterworth-Heinemann
An imprint of Elsevier

Butterworth-Heinemann is an imprint of Elsevier
The Boulevard, Langford Lane, Kidlington, Oxford OX5 1GB, United Kingdom
50 Hampshire Street, 5th Floor, Cambridge, MA 02139, United States

Copyright © 2018 Elsevier Inc. All rights reserved

No part of this publication may be reproduced or transmitted in any form or by any means,
electronic or mechanical, including photocopying, recording, or any information storage and
retrieval system, without permission in writing from the publisher. Details on how to seek
permission, further information about the Publisher's permissions policies and our arrangements
with organizations such as the Copyright Clearance Center and the Copyright Licensing Agency,
can be found at our website: www.elsevier.com/permissions.

This book and the individual contributions contained in it are protected under copyright by the
Publisher (other than as may be noted herein).

Notices

Knowledge and best practice in this field are constantly changing. As new research and experience
broaden our understanding, changes in research methods, professional practices, or medical
treatment may become necessary.

Practitioners and researchers must always rely on their own experience and knowledge in evaluating
and using any information, methods, compounds, or experiments described herein. In using such
information or methods they should be mindful of their own safety and the safety of others,
including parties for whom they have a professional responsibility.

To the fullest extent of the law, neither the Publisher nor the authors, contributors, or editors,
assume any liability for any injury and/or damage to persons or property as a matter of products
liability, negligence or otherwise, or from any use or operation of any methods, products,
instructions, or ideas contained in the material herein.

Library of Congress Cataloging-in-Publication Data
A catalog record for this book is available from the Library of Congress

British Library Cataloguing-in-Publication Data
A catalogue record for this book is available from the British Library

ISBN: 978-0-12-813788-8

For information on all Butterworth-Heinemann publications
visit our website at https://www.elsevier.com/books-and-journals

Working together
to grow libraries in
developing countries

www.elsevier.com • www.bookaid.org

Publisher: Marr Conner
Acquisition Editor: Sonnini R. Yura
Editorial Project Manager: Natasha Welford
Production Project Manager: Anitha Sivaraj
Designer: Christian Bilbow

Typeset by VTeX

The first author dedicates this book to her husband, Gilberto, her mother Yolanda, and her children: Alma Sofia and Daniela Monserrat.

The second author dedicates this book to her husband, Angel, her children Ana, Sara, and Angel, as well as her parents Maria and Trinidad, and her brothers and sisters: Rodolfo, Claudia, Nora, Carlos, Ernesto, Gerardo, and Paola.

The third author dedicates this book to his wife, Paty, and his children: Carlos Alejandro, Fernando Yhael, and Íker Mateo.

CONTENTS

PREFACE

Bio-inspired algorithms have become an important research area due to intent to emulate nature in order to help solve real-life complex problems, particularly for optimization. Due to the high level of enthusiasm generated by successful applications, the use of bio-inspired algorithms to solve complex optimization problems in a heuristic way has become a well-established methodology. This book proposes novel algorithms, including combined well-known bio-inspired algorithms, to solve real-life complex problems in classification, approximation, vision, pattern recognition, identification, and control. Rigorous analyses as well as unique applications of these algorithms are also presented.

Most research in this field has two main focuses: one dedicated to solve academic problems (complex but only with academic meaning) and the second, to develop more bio-inspired algorithms. The work is intended to alleviate well-known problems such as slow convergence, stagnation in local minimal, and high computational complexity, among others.

In this book authors present a set of real-life bio-inspired algorithms applications, including intelligent pattern recognition, object reconstruction, robot control and vision, intelligent identification, and control of nonlinear systems. Nevertheless, the main drawback of this book is that the proposed methodologies are designed to deal with a wide range of engineering problems and not limited to applications selected to show their effectiveness. The proposed applications of each methodology, however, is at the level of state of the art of their respective field. The wide range of considered applications shows the capacity of bio-inspired algorithms to solve real-life problems.

The main goal of this book is to facilitate the application of many proposed bio-inspired algorithms developed in last few decades to real-life problems and not limit the same to an academic context. To achieve this, the book covers both theoretical and practical methodologies to allow readers greater appreciation regarding the implementation of bio-inspired algorithms.

The vast majority of other books related to bio-inspired algorithms and a great number of scientific papers only show the respective algorithms, pseudo-codes, flux diagrams, etc. but do not include real-life problems; they are limited to showing the effectiveness of bio-inspired algorithms to solve academic problems, typically benchmark functions. In this book the algorithms are presented in a rather friendly manner giving more emphasis to real-life applications and their implementation without neglecting their mathematical foundations, including both simulation and experimental results. The book also contains rigorous analysis of the proposed and/or used bio-inspired algorithms in addition to fundamental aspects of the proposed topics, development, variants, and modifications.

This book is organized as follows:

In Chapter 1 bio-inspired algorithms are introduced, and the algorithms used in the book are presented. The latter include Particle Swarm Optimization (PSO), Artificial Bee Colony Algorithm (ABC), Micro Artificial Bee Colony Algorithm (μABC), Differential Evolution (DE), and Bacterial Foraging Optimization Algorithm (BFO).

In Chapter 2 an approach to large-data classification using support vector machines trained with evolutionary algorithms employing Kernel Adatron are introduced. Basic concepts of support vector machines are conveyed, and Kernel Adatron definition, its training with bio-inspired algorithms and its application to different benchmark repository and to classify electromyographic signals presented.

In Chapter 3 reconstruction of 3D surfaces using radial basis functions adjusted with PSO is described, including foundations of interpolation and its application to 3D surface approximation of point-clouds.

Chapter 4 shows soft computing applications in robot vision and carries a brief introduction to robot vision, image tracking, plane detection, pose estimation, and their application to a real-life scenario.

Chapter 5 includes soft computing applications in mobile robotics and provides an introduction to mobile robotics, holonomic, and nonholonomic mobile robot application.

Chapter 6 deals with applying PSO to improve neural identifiers for discrete-time unknown nonlinear systems applied to the model in a first stage, subsequently for the reestablishment of a linear induction motor benchmark, and then to produce a nonlinear model for forecasting in smart grids.

Chapter 7 proposes the use of bio-inspired algorithms to improve neural controllers for discrete-time unknown nonlinear systems particularly with the use of PSO and BFO with a Neural Second-Order Sliding Mode Controller for trajectory tracking for a large class of discrete-time unknown nonlinear systems.

Finally, relevant conclusions and future work are stated in Chapter 8.

This book could be used for self-learning as well as a textbook. Our target audience includes, but is not limited to professors, research engineers, and graduate students carrying out work in the areas of artificial intelligence, robotics, machine learning, vision, classification, pattern recognition, identification and control, among others. However, due to bio-inspired algorithms having become a well-established research area with many applications in different scientific fields, ranging from science and engineering to economics and social sciences, it is not possible to restrict the scope that this book can have regarding the possible applications of the methodologies presented herein.

Alma Y. Alanis
Nancy Arana-Daniel
Carlos López-Franco
Guadalajara, Jalisco, Mexico
August 2017

ACKNOWLEDGMENTS

The authors thank CONACYT (Spanish acronym, which stands for National Council of Sciences and Technology), Mexico, for financially supporting the following projects: CB-256769, CB-256880, and PN-2016-4107. They also thank CUCEI-UDG (Spanish acronym for University Center of Exact Sciences and Engineering of the University of Guadalajara), Mexico, for the support provided to write this book. The first author also offers thanks for the support given by "Fundacion Marcos Moshinsky."

In addition, gratitude is extended to Luis J. Ricalde, Eduardo Rangel, Chiara Simetti, Jorge Rios, Carlos Villaseñor, Alejandro Gallegos, Gehová López, Roberto Valencia, Jacob Morales, José de Jesús Hernández, and Javier Gómez, who have contributed in different ways to the composition of this book.

CHAPTER ONE

Bio-inspired Algorithms

Contents

1.1. INTRODUCTION

In the last few decades, bio-inspired algorithms have become a well-established research area with many applications in different scientific fields, ranging from science and engineering to economics, and the social sciences. However, most of such works can be divided into two main sets. The first set proposes new bio-inspired algorithms with important characteristics to solve complex optimization problems, and the second set includes their applications. Nevertheless, most of such works only deal with academic examples. In literature, there are few real-life applications, and less new proposals to deal with real-life problems.

Bio-inspired algorithms have become an important research area due to intent to emulate nature in order to help us solve real-life complex problems, particularly for optimization. Bio-inspired algorithms for engineering propose novel algorithms to solve real-life complex problems in classification, approximation, vision, pattern recognition, identification, and control. Such algorithms include combined well-known bio-inspired algorithms and new ones, including both rigorous analyses, as well as unique applications.

From the beginning of the computation era, people have been trying to optimize the processes by which a particular program solves a task entrusted; from the optimization of resources to the execution time. With the increasing popularity of computer systems came increasingly complex tasks for which conventional algorithms were inefficient, so it became necessary

Bio-inspired Algorithms for Engineering
https://doi.org/10.1016/B978-0-12-813788-8.00001-9
© 2018 Elsevier Inc.
All rights reserved.
1

to provide some kind of intelligence to the algorithms. This particular form of intelligence was named artificial intelligence.

In order to try to define the so-called artificial intelligence, it is necessary to describe the characteristics that until now are known to build the intelligence, and then to extrapolate them to the artificial processes (algorithms) that will use it. It is important to note that these characteristics are not absolute, nor independent, but are combined synergistically to give way to the term "intelligent". Some of the intelligence features—in terms of optimization—are adaptability, randomness, communication, feedback, exploration, and exploitation. It is important to mention that although all these characteristics are necessary to denote intelligence, none of them is sufficient, which is why there is a relationship of cooperation rather than competition [12].

Bio-inspired systems are linked to several areas whether within an industrial, scientific, or even social or natural disciplines. In determining one, or several possible solutions to a given problem, it tends to be updatable through all the interactions to which it is immersed, that is, it always evolves. There are several terms that are associated with bio-inspired algorithms, and the most representative are:

- evolutionary computing,
- population-based optimization,
- computational intelligence,
- soft computing,
- nature-inspired computing,
- machine learning,
- heuristic algorithms, and
- collective intelligence.

While all these terms have some relation, there are also very subtle differences between them, so the terminology is imprecise and highly context-dependent. However, this book will use the term bio-inspired algorithm. Typically, an iteration of a bio-inspired algorithm is called a generation, this to maintain its biological foundation. So, the terminology in bio-inspired algorithm is not uniform and may be confusing, mainly because it is a relatively young research area, in constant growth and has not yet reach its maturity [12]. It is well known that bio-inspired algorithms closely follow the theory of natural evolution, which rests on four pillars: population, diversity, heredity, and selection. However, biology is making continuous progress in the description of the components that make up living organisms and of the ways in which those components work together [6].

The concept of evolution did not exist formally until Charles Darwin published his book entitled "Origin of Species" in 1859. From that moment was created the theory that would base the genetics whose origin is attributed to Gregor Mendel, who in 1865 published his findings in his report "Experiments on hybridization of plants". Therein he conveyed the mechanisms of natural selection what later became known as the Law of the inheritance of Mendel. These concepts along with those specifically related to DNA, genes, chromosomes, mutation, fitness, and crossover built the concept of bio-inspired algorithms.

1.1.1 Bio-inspired and evolutionary algorithms

Evolutionary computing is a subfield of artificial intelligence that includes a range of problem-solving techniques based on principles of biological evolution. The principles for using evolutive processes to solve optimization problems originated in the 1950s [5].

The EA are optimization methods that are part of evolutionary computing, applying models based on biological evolution. In EA, a population of possible solutions is composed of individuals that can be compared according to their aptitude to improve the population; the most qualified candidates are those that obtain better results by a fitness function evaluation. The evolution of the population is obtained through iterations, in which a series of operations are applied to the individuals of the population (reproduction, mutation, recombination or selection). From these operations a new set of potentially better solutions are generated. The way the population evolves the possible solutions, and the way it chooses the new global best solutions, is something inherent to each EA [12].

A swarm intelligence algorithm is based on swarms that occur in nature; PSO and ABC are two prominent swarm algorithms. There is a debate on whether swarm intelligence-based algorithms are EAs or not, but since one of the inventors of PSO refers to it as an EA, and swarm intelligence algorithms are executed in the same general way as EAs—evolving a population of candidate problem solutions that improves with each iteration—we consider swarm intelligence to be an EA [12,11].

The rest of this chapter is devoted to explain EAs as used throughout this book. These include PSO, ABC, μABC, DE, and BFO, all used to solve different engineering problems such as data classification, reconstruction of 3D surfaces, robot vision, mobile robotics, neural identification, and neural control design.

1.2. PARTICLE SWARM OPTIMIZATION

The PSO algorithm was first introduced by Kennedy and Russell in 1995 [8]. This algorithm exploits a population of potential solutions. The population of solutions is called a swarm, and each individual from a swarm is called a particle. A swarm is defined as a set of n particles. Each particle i is represented as a D-dimensional position vector x_i, which is evaluated by a fitness function $f(\cdot)$. Based on the results of the evaluation, it is easy to measure improvement in new particles compared to old ones. The particles are assumed to move within the search space iteratively. This is done by adjusting their position using a proper position shift, called velocity v_i.

For each iteration t, the velocity changes by applying Equation (1.1) to each particle.

$$v_i(t+1) = \omega v_i(t) + c_1 \varphi_1 (P_{ibest} - x_i) + c_2 \varphi_2 (P_{gbest} - x_i), \qquad (1.1)$$

where φ_1 and φ_2 are random variables uniformly distributed within $[0, 1]$; c_1 and c_2 are weighting factors, also called the cognitive and social parameters, respectively; ω is called the inertia weight, which decreases linearly from ω_{start} to ω_{end} during iterations; and P_{ibest} and P_{gbest} represent the best position visited by a particle and the best position visited by the swarm before the current iteration t, respectively.

The position update is applied by Equation (1.2) based on the new velocity and the current position.

$$x_i(t+1) = x_i(t) + v_i(t+1). \qquad (1.2)$$

The basic algorithm is Algorithm 1.

To solve the uncontrolled increase of magnitude of the velocities (swarm explosion effect), it is often necessary to restrict the velocity with a clamping at desirable levels, preventing particles from taking extremely large steps from their current positions [9]:

$$v_{ij}(t+1) = \begin{cases} v_{max} & \text{if } v_{ij}(t+1) > v_{max}, \\ -v_{max} & \text{if } v_{ij}(t+1) < -v_{max} \end{cases}$$

Although the use of a maximum velocity threshold improves the performance by controlling the swarm explosions, without the inertia weight, the swarm would not be able to concentrate its particles around the most promising solutions in the last phase of the optimization procedures [9].

Algorithm 1 Particle Swarm Optimization.
1: Initialize c_1, c_2, v_i and x_i
2: $P_{ibest} \leftarrow x_i$.
3: Select from x_i, P_{gbest}.
4: **repeat**
5:    Obtain velocity v_i with Equation (1.1).
6:    Update position x_i with Equation (1.2).
7:    **if** $f(x_i) < f(P_{ibest})$ **then**
8:       $P_{ibest} \leftarrow x_i$
9:       **if** $f(P_{ibest}) < f(P_{gbest})$ **then**
10:          $P_{gbest} \leftarrow P_{ibest}$
11:       **end if**
12:    **end if**
13: **until** The stopping condition is met.

1.3. ARTIFICIAL BEE COLONY ALGORITHM

The ABC algorithm was first introduced by Karaboga [7] in 2005. This algorithm is based on the honey bee foraging behavior. The bees are divided in three classes:

- employed: bee with a food source;
- onlookers: bee that watches the dances of employed bees and choose food sources depending on dances;
- scouts: employed bee that abandons its food source to find a new one.

Each food source is equivalent to a possible solution to the optimization problem and, as in nature, individuals are more likely to be attracted to sources with a larger amount of food (a better result obtained by the fitness function).

For each food source, only one employed bee is assigned, and when it abandons its food source, it becomes a scout. The number of the onlooker bees is also equal to the number of solutions in the population.

Initially, ABC algorithm generates a random population P of n solutions, where each solution $x_i \in P$ is an n-dimensional vector. The algorithm searches iteratively for the better food sources based on the findings made by employed, onlooker, and scout bees. First, the ith employed bee produces a random modification on x_{ij} in its jth position (where $1 \leq j \leq D$), and it takes it as a new potential food source v_i. The potential food source can be obtained by the equation

$$v_{ij} = x_{ij} + \phi_{ij}(x_{ij} - x_{kj}), \tag{1.3}$$

where $k \in 1, 2, ..., n$ is a randomly chosen index different from i, and ϕ_{ij} is a uniformly distributed number between $[-1, 1]$.

If the amount of nectar (the value obtained by the fitness function) is greater than the old one, then the employed bee takes it as the new position x_i. Otherwise the food source x_i remains unchanged.

Once the updates on the positions have been made by the employed bees, the information is shared with the onlooker bees. Onlooker bees choose their food sources based on a probability p_i that is directly related to the amount of nectar. The value of p_i is obtained as follows:

$$p_i = \frac{f_i}{\sum\limits_{m=1}^{n} f_m}, \tag{1.4}$$

where f_i is the fitness value of the ith food source. p_i is chosen by a roulette wheel selection mechanism (the better the ith solution, the higher its chances of being selected).

A new potential food source v_i is calculated using Equation (1.3), where x_{ij} is selected based on the roulette wheel selection result, and, as with employed bees, if the amount of nectar improves, then v_i replaces x_i; otherwise, x_i remains unchanged.

If a position x_i cannot be improved through a certain number of iterations It, then the ith food source is abandoned. If this occurs, then the scout bee changes its actual food source for a new food source to replace x_i as follows:

$$x_{ij} = lb_j + rand(0, 1)(ub_j - lb_j), \tag{1.5}$$

where $rand(0, 1)$ is a normally distributed random number within $[0, 1]$, and lb and ub are lower and upper bounds of the jth dimension, respectively. The pseudocode is briefly described in Algorithm 2.

1.4. MICRO ARTIFICIAL BEE COLONY ALGORITHM

The μABC algorithm was first introduced by Rajasekhar in 2012 [10]. This algorithm is a variant of the ABC algorithm with a small population (only three bees). The population of bees evolves through iterations,

---

Algorithm 2 Artificial Bee Colony Algorithm.

1: Initialize x_i
2: **repeat**
3:     Produce a new solution v_i for the employed phase with Eq. (1.3).
4:     **if** $f(v_i) < f(x_i)$ **then**
5:         $x_i \leftarrow v_i$.
6:     **end if**
7:     Calculate the probability values pi with Eq. (1.4) for the solution x_i.
8:     Produce a new solution v_i for the onlooker phase with Eq. (1.3), selecting x_i based on p_i.
9:     **if** $f(u_i) < f(x_i)$ **then**
10:       $x_i \leftarrow v_i$.
11:     **end if**
12:     **if** x_i is an abandoned solution for the scout phase **then**
13:       Replace x_i by using the Eq. (1.5).
14:     **end if**
15: **until** The stopping condition is met.

---

and only the best bee is kept unaltered, whereas the rest of the bees are reinitialized with modifications based on the food source with the best fitness.

After the employed and onlooker phases have been completed (in the same way that are completed in the ABC algorithm), the population is ranked in accordance to its fitness values. The bee with the best fitness remains in its food source, whereas the second best fitness is moved to a position near to the best one in order to facilitate local search. The bee with the worst position is initialized to a random position to avoid premature convergence.

Unlike ABC, more than one variable is modified from the food source. For each parameter x_{ij}, a uniformly distributed random number $rand_{ij}(0,1)$ is generated, and if this number is less than the Frequency Control Rate (FCR) parameter, then the variable x_{ij} is modified as follows:

$$v_{ij} = \begin{cases} x_{ij} + \phi_{ij}(x_{ij} - x_{kj}) & \text{if } rand(0,1) \leq FCR, \\ x_{ij} & \text{otherwise.} \end{cases} \tag{1.6}$$

The value of ϕ_{ij} is a uniformly distributed random number, maintained in the range of $[-RF, RF]$, where RF is the range factor. RF changes automatically during the search by tuning its value in accordance to the Rechenberg 1/5 rule. This rule states that 1/5 of the total mutations every t iterations $\varphi(t)$ should be successful mutations. According to the number of successes $\varphi(t)$, the value of RF is adjusted according to

$$RF_{new}(it + 1) = \begin{cases} RF_{old}(t) * 0.85 & \text{if } \varphi(t) < 1/5, \\ RF_{old}(t)/0.85 & \text{if } \varphi(t) > 1/5, \\ RF_{old}(t) & \text{if } \varphi(t) = 1/5. \end{cases} \qquad (1.7)$$

The pseudocode is briefly described in Algorithm 3.

Algorithm 3 Micro Artificial Bee Colony Algorithm.

1: Initialize x_i
2: **repeat**
3:     Produce a new solution v_i for the employed phase with Eq. (1.6).
4:     **if** $f(v_i) < f(x_i)$ **then**
5:         $x_i \leftarrow v_i$.
6:     **end if**
7:     Calculate probability values pi with Eq. (1.4) for solution x_i.
8:     Produce a new solution v_i for the onlooker phase with Eq. (1.6), selecting x_i based on p_i.
9:     **if** $f(u_i) < f(x_i)$ **then**
10:         $x_i \leftarrow v_i$.
11:     **end if**
12:     Move second best solution x_{2b} to a position very close to best solution x_{1b}.
13:     Move worst solution x_{3b} to a random position.
14: **until** The stopping condition is met.

1.5. DIFFERENTIAL EVOLUTION

In Differential Evolution (DE) each individual x_i of the population is an n-dimensional vector that represents a candidate solution from a set of N solutions. The fundamental idea behind DE is creating new candidate solutions based on other solutions that had been previously found. DE takes the difference vector between two randomly chosen individuals, x_{r2} and x_{r3}, and adding a scaled version of this vector to a third individual, whom could be chosen randomly x_{r1} or by selecting the best individual x_b in the population. For the algorithm described in this chapter, we used $x_{r1} = x_b$. This new individual is called a mutant vector v_i:

$$v_i = x_{r1} + F(x_{r2} - x_{r3}), \tag{1.8}$$

where F is a scaling factor. This mutant vector v_i is later combined with x_i by crossover to create a candidate solution to be evaluated by an objective function. The crossover is implemented as follows:

$$u_{ij} = \begin{cases} v_{ij} & \text{if } (r_{cj} < CROV) \text{ or } (j = J_r), \\ x_{ij} & \text{otherwise} \end{cases} \tag{1.9}$$

$$\text{for } i = 1, 2, ..., N \text{ and } j = 1, 2, ..., n,$$

where u_{ij} is the crossed vector, r_{cj} is a random number between $[0, 1]$, $CROV$ is the constant crossover rate $\in [0, 1]$, and J_r is a random integer $\in [0, 1]$. The pseudocode is briefly described in Algorithm 4.

Algorithm 4 Differential Evolution Algorithm.

1: Initialize $F = [0.4, 0.9]$, $c = [0.1, 1]$ and x_i
2: **repeat**
3:     For each x_i chose three random integers $(r1, r2, r3)$, where $r1 \neq r2 \neq r3$ and $r1, r2, r3 \in [1, N]$.
4:     Generate N mutant vectors with Eq. (1.8).
5:     Generate N crossed vectors with Eq. (1.9).
6:     **if** $f(u_i) < f(x_i)$ **then**
7:         $x_i \leftarrow u_i$.
8:     **end if**
9: **until** The stopping condition is met.

1.6. BACTERIAL FORAGING OPTIMIZATION ALGORITHM

The Bacterial Foraging Optimization (BFO) algorithm was first proposed in 2002 by Passino. It is inspired by the foraging behavior and chemotactic bacteria, especially *Escherichia coli* (*E. coli*) in our intestine. By running smooth and tumbling, it can be moved to the escape area of nutrients and poison zone in the environment. Chemotaxis is more attractive behavior of bacteria, and it has been studied by many researchers [1,3,4,2]. Due to their group response, the social behavior of the colony of *E. coli* is very interesting for engineering; it allows them to get quickly and easily the best food supply with the lowest possible risk. These bacteria can communicate through chemical exchanges. The bacteria that have achieved a safe place to feed, communicate it to others who come to such place; the greater the amount of existing food, the stronger is the signal. Similarly, if the bacteria are in a dangerous place, with agents that may threaten the colony, they warn others to stay away from that place. This behavior can be represented mathematically forage as a kind of swarm intelligence [1].

So, BFO mimics the four principal mechanisms of real bacterial systems: chemotaxis, swarming, reproduction and elimination-dispersal to solve optimization problems [2]. The pseudocode of the algorithm is shown in Algorithm 5.

- **Chemotaxis:** This process simulates the movement of an *E. coli* bacterium through swimming and tumbling via flagella. Biologically an *E. coli* bacterium can move in two different ways: it can swim for a period of time in the same direction, or it may tumble, and alternate between these two modes of operation for its entire lifetime.

- **Swarming:** It is a behavior that has been observed for several motile species of bacteria, including *E. coli*, where intricate and stable spatiotemporal swarms are formed in semisolid nutrient medium. A group of *E. coli* cells arrange themselves in a traveling ring by moving up the nutrient gradient when placed amidst a semisolid matrix with a single nutrient chemo-effecter. The cells when stimulated by a high level of *succinate*, release an attractant, *aspartate*, which helps them to aggregate into groups and thus move as concentric patterns of swarms with high bacterial density. The cell-to-cell signaling in *E. coli* swarm may be represented by the following function:

$$J_{cc}(\theta, P(j, k, l)) = \sum_{i=1}^{S} J_{cc}(\theta, \theta^i(j, k, l))$$

$$= \sum_{i=1}^{S} \left[-d_{attractant} exp(-w_{attractant} \sum_{m=1}^{p} (\theta_m - \theta_m^i)^2) \right]$$

$$+ \sum_{i=1}^{S} \left[h_{repellant} exp(-w_{repellant} \sum_{m=1}^{p} (\theta_m - \theta_m^i)^2) \right], \quad (1.10)$$

where $J_{cc}(\theta, P(j, k, l))$ is the objective function value to be added to the actual objective function to present a time varying objective function, $\theta = [\theta_1, \theta_2, ..., \theta_p]^T$ is a point in the p-dimensional search domain, and $d_{attractant}, w_{attractant}, h_{repellant}, w_{repellant}$ are different coefficients that have to be chosen [3].

- **Reproduction:** The least healthy bacteria eventually dies while each of the healthier bacteria (those yielding lower value of the objective function) asexually split into two bacteria, which are then placed in the same location. This keeps the swarm size constant.

- **Elimination and Dispersal:** Gradual or sudden changes in the local environment where a bacterium population lives may occur due to various reasons, e.g. a significant local rise of temperature may kill a group of bacteria that are currently in a region with a high concentration of nutrient gradients. Events can take place in such a fashion that all the bacteria in a region are killed or a group is dispersed into a new location. To simulate this phenomenon in BFOA, some bacteria are liquidated at random with a very small probability while the new replacements are randomly initialized over the search space.

Next, it is going to be defined when a chemotactic step is a tumble followed by a tumble or a tumble followed by a run. Consider the next variables' definitions:

- j, index for the chemotactic step;
- k, index for the reproduction step;
- l, index of the elimination-dispersal event;
- p, dimension of the search space;
- S, total number of bacteria in the population;
- N_C, number of chemotactic steps;
- N_s, swimming length;
- N_{re}, number of reproduction steps;
- N_{ed}, number of elimination-dispersal events;
- P_{ed}, elimination-dispersal probability;
- $C(i)$, size of the step taken in the random direction specified by the tumble, and
- θ^i, ith bacterium.

Algorithm 5 Bacterial Foraging Optimization Algorithm.

1: Initialize parameters $p, S, N_C, N_s, N_{re}, N_{ed}, P_{ed}C(i)$ for $i = 1, 2, ..., S$, θ^i

2: Elimination-dispersal loop: $l = l + 1$

3: Reproduction loop $k = k + 1$

4: Chemotaxis loop $j = j + 1$

5: **for** $i = 1, 2, ...S$ **do**

6:     Take a chemotactic step for bacterium i

7:     Compute fitness function $J(i, j, k, l) = J(i, j, k, l) + J_{cc}(\theta^i(j, k, l), P(j, k, l))$, see Eq. (1.10)

8:     $J_{last} = J(i, j, k, l)$

9:     Tumble: generate a random vector $\Delta(i) \in \Re^p$ with each element $\Delta_m(i), m = 1, 2, ..., p$ a random number on $[-1, 1]$

10:     Move: Let $\theta^i(j + 1, k, l) = \theta^i(j, k, l) + C(i)\frac{\Delta(i)}{\sqrt{\Delta^T(i)\Delta(i)}}$

11:     Compute $J(i, j + 1, k, l) = J(i, j, k, l) + J_{cc}(\theta^i(j + 1, k, l), P(j + 1, k, l))$

12:     Let $m = 0$ (counter for swim length)

13:     **while** $m < N_S$ **do**

14:         Let $m = m + 1$

15:         **if** $J(i, j + 1, k, l) < J last$ **then**

16:             Let $J_{last} = J(i, j + 1, k, l)$

17:             Let $\theta^i(j + 1, k, l) = \theta^i(j, k, l) + C(i)\frac{\Delta(i)}{\sqrt{\Delta^T(i)\Delta(i)}}$

18:             Compute $J(i, j + 1, k, l) = J(i, j, k, l) + J_{cc}(\theta^i(j + 1, k, l), P(j + 1, k, l))$

19:         **else**

20:             Let $m = N_S$

21:         **end if**

22:     **end while**

23:     Go to the next bacterium $i + 1$, if $i \neq S$

24: **end for**

25: If $j < N_c$ go to chemotaxis loop. In this case continue chemotaxis since the life of the bacteria is not over.

26: Reproduction:

27: **for** $i = 1, 2, ..., S$ and for the given k and l **do**

28:     Let $J_{health}^i = \sum_{j=1}^{N_c+1} J(i, j, k, l)$ be health for the bacterium i

29:     Sort bacteria and chemotactic parameters $C(i)$ in order of ascending cost J_{health}

30: **end for**

Algorithm 5 *continued*

31: The S_r bacteria with the highest J_{health} values die, and the remaining S_r bacteria with the best values split.

32: If $k < N_{re}$ go to step 3. In this case, we have not reached the number of specified reproduction steps, so we start the next generation of chemotactic loop.

33: Elimination-dispersal:

34: **for** $i = 1, 2, ..., S$ with probability P_{ed} **do**

35:     Eliminate and disperse each bacterium. If a bacterium is eliminated, simply disperse another one to a random location on the optimization domain.

36:     **if** $l < N_{ed}$ **then**

37:         Go to Elimination-dispersal loop

38:     **else**

39:         End the algorithm iteration

40:     **end if**

41: **end for**

REFERENCES

[1] Bhushan B, Singh M. Adaptive control of nonlinear systems using bacterial foraging algorithm. Int J Comput Electr Eng 2011;3(3):335.

[2] Das S, Biswas A, Dasgupta S, Abraham A. Bacterial foraging optimization algorithm: theoretical foundations, analysis, and applications. In: Abraham A, Hassanien AE, Siarry P, Engelbrecht AP, editors. Foundations of computational intelligence (3). Studies in Computational Intelligence, vol. 203. Springer; 2009. p. 23–55. [Online]. Available from: http://dblp.uni-trier.de/db/series/sci/sci203.html#DasBDA09.

[3] Das S, Dasgupta S, Biswas A, Abraham A, Konar A. On stability of the chemotactic dynamics in bacterial-foraging optimization algorithm. Trans Syst Man Cyber Part A May 2009;39(3):670–9. [Online]. Available from: https://doi.org/10.1109/TSMCA.2008.2011474.

[4] Dasgupta S, Das S, Abraham A, Biswas A. Adaptive computational chemotaxis in bacterial foraging optimization: an analysis. Trans Evol Comput Aug. 2009;13(4):919–41. [Online]. Available from: https://doi.org/10.1109/TEVC.2009.2021982.

[5] Dianati M, Song I, Treiber M. An introduction to genetic algorithms and evolution strategies. Tech. rep. Ontario, N2L 3G1, Canada: University of Waterloo; 2002.

[6] Floreano D, Mattiussi C. Bio-inspired artificial intelligence: theories, methods, and technologies. The MIT Press; 2008.

[7] Karaboga D. An idea based on honey bee swarm for numerical optimization. Tech. rep. Engineering Faculty, Computer Engineering Department; 2005.

[8] Kennedy J, Eberhart R. Particle swarm optimization. In: Proceedings of IEEE international conference on neural networks, vol. 4. IEEE; November 1995. p. 1942–8.

[9] Parsopoulos K, Vrahatis M. Particle swarm optimization and intelligence: advances and applications. IGI Global; 2010.

[10] Rajasekhar A, Das S, Das S. μABC: a micro artificial bee colony algorithm for large-scale global optimization. In: GECCO'12 companion. ACM; 2012. p. 1399–400.

[11] Shi Y, Eberhart R. Empirical study of particle swarm optimization. In: Proceedings of the 1999 congress on evolutionary computation, vol. 3; 1999. p. 1950.

[12] Simon D. Evolutionary optimization algorithms: biologically-inspired and population-based approaches to computer intelligence. John Wiley & Sons, Inc.; 2013.

CHAPTER TWO

Data Classification Using Support Vector Machines Trained with Evolutionary Algorithms Employing Kernel Adatron

Contents

2.1. INTRODUCTION

With the drastic increase of data in a rather short length of time, new problems have appeared. Problems like classifying chromosomes, spam filtering, defining which advertisements to show to a person in a web page, etc. [25,13], generate an immense amount of data. Sometimes the data are so big that the potential size of training sets cannot be stored in a computer. This problem is solved by large-scale learning which aims to find a function that converts the data, and their corresponding class labels, to an amount of data that can be stored in a modern computer's memory [9]. The main constraint is the length of time that an algorithm takes to obtain an accurate result rather than the number of samples to process [30].

A typical problem that Support Vector Machines (SVM) have to face while working with large dataset is that learning algorithms are typically quadratic and require several scans of the data. To solve this, three common strategies can be distinguished to reduce this practical complexity [9,30]:

- solving several smaller problems by working on subsets of the training data instead of the complete large dataset,

Bio-inspired Algorithms for Engineering
https://doi.org/10.1016/B978-0-12-813788-8.00002-0

© 2018 Elsevier Inc.
All rights reserved.

- parallelizing the learning algorithm, and
- designing a less complex algorithm that give an approximate solution with equivalent or superior performance.

This work presents a novel approach to solving large-scale learning problems by designing a less complex algorithm to train a large-scale SVM. Our approach uses a combination of Kernel-Adatron (KA) and some state-of-the-art evolutionary algorithms (EAs) to solve large-scale learning problems [17]. The obtained algorithm works with small subproblems, has low computational complexity, and is easy to implement. In addition to providing accurate generalization results, such methodology is also highly parallelizable.

2.2. SUPPORT VECTOR MACHINES

Since SVM was first introduced by Vladimir Vapnik in 1995 [10], it has been one of the most popular methods for classification due to their characteristics. Such characteristics include its simple model, the use of kernel functions, and the convexity of the function to optimize (it only has a global minimum). SVM's characteristics make it more appealing compared to the Multilayer Perceptron [35]. Many large-scale training algorithms have been proposed for SVMs [28,39,42,15]. The main idea of these algorithms is to minimize a regularized risk function R and maximize the margin of separation between classes (Fig. 2.1) by solving the equation

$$w^* = \arg\min_{w \in \Re^D} F(w) := \frac{1}{2} \|w\|^2 + CR(w), \tag{2.1}$$

where w is a normal vector, $\frac{1}{2} \|w\|^2$ is a quadratic regularization term, and $C > 0$ is the fixed constant that scales the risk function.

Equation (2.1) is called the primal formulation [19]. By using Lagrange multipliers the primal formulation can be presented in its dual form [30]:

$$L(\alpha) = \arg\max_{\alpha} \{\sum_{i=1}^{n} \alpha_i - \sum_{i=1}^{n} \sum_{j=1}^{n} \alpha_i \alpha_j \gamma_i \gamma_j K(x_i, x_j)\}$$

$$\text{subject to } 0 \leq \alpha_i \leq C \text{ and } \sum_{i=1}^{n} \alpha_i \gamma_i = 0, \tag{2.2}$$

where C is a fixed constant, $(X_i, Y_i)_{i=1}^{n}$ is a training set, α_i are Lagrange multipliers, $K(X_i, X_j)$ is the value of the kernel matrix defined by the inner

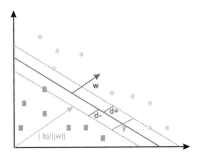

Figure 2.1 For purposes of generalization, the hyperplane with the largest margin $(d_+ + d_-)$ gives the best results, although there can be several hyperplanes that optimally separate a dataset.

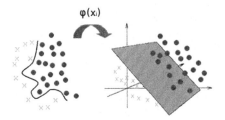

Figure 2.2 Data sets that are not linearly separable can be separated by a hyperplane in higher dimensions after applying the kernel trick.

product $\langle X_i, X_j \rangle$ (when a linear kernel K is used), and $Y_i \in \{\pm 1\}$ is a class label [30].

The dual formulation has the same optimal values as the primal, but the main advantage of this representation is the use of the "kernel trick" (see Fig. 2.2). Since SVMs can only classify data in a linear separable feature space, the role of the kernel-function is to induce such feature space by implicitly mapping the training data into a higher dimensional space where data is linear separable [19,32].

There are two main sets of approaches for large-scale SVM training algorithms: those that solve the primal SVM formulation, shown in Eq. (2.1) by a gradient-based method (primal estimated subgradient solver for SVM, careful quasi-Newton stochastic gradient descent, forward looking subgradient, etc.), and those that solve the dual formulation of Eq. (2.2) by Quadratic Programming (QP) methods (SVM for multivariate performance measure, library for large linear classification, and bundle method for risk minimization, etc.) [30,39,40,8]. There are options that do not fall into these categories, such as the optimized cutting plane algorithm (OCA), which uses an improved cutting plane technique, based on the work of SVM for multivariate performance measure (SVM^{perf}), and bundle method for risk minimization. OCA has fast convergence compared to methods like stochastic gradient descent and primal estimated subgradient solver for SVM (Pegasos), and it has shown good classification results and offers computational sublinear scaling [15]. Nevertheless, the use of a QP solver to solve a linear constraint problem (where each linear constraint is a cutting plane) makes it a complex approach to implement, even if the number of constraints is drastically lower than the data dimensionality.

Gradient-based methods tend to be fast algorithms (especially those that use stochastic gradient descent), and have good generalization capabilities. However, they are highly dependent on step size to obtain a good speed of convergence. If the step size is not chosen carefully or it does not have an adjustment criterion, this can produce slow convergence [30]. The dual QP methods can handle kernels easily and can converge quickly by combining them with other optimization techniques. The main disadvantage of these methods is the computational complexity of the quadratic programming solvers and the fact that they are more difficult to implement than a gradient descent method, or an EA [30,43,18,14,16].

In past years, several evolutionary computation-based training algorithms for SVM have been proposed [31,29,1,20]. These algorithms solve the dual formulation (Eq. (2.2)), tend to be easy to implement and have shown good results for small amounts of data. The disadvantage of their implementation is their computational complexity of $O(n^2)$ or higher, where n represents the number of training samples. Since the complete kernel is needed on each iteration to calculate the fitness function, as the number of training samples grows, the time needed to process the data will increase drastically.

2.3. EVOLUTIONARY ALGORITHMS

As we stated in Chapter 1, EAs are global optimization methods that scale well to higher-dimensional problems. They are robust with respect to noisy evaluation functions, and can be implemented and parallelized with relative ease [41]. Even when premature convergence to a local extremum may occur, it has been proven that an algorithm that is "not quite good," or "poor" at optimization, can be excellent at generalization for a large-scale learning task [30].

This work presents a series of parallelized algorithms based on the KA algorithm as fitness function, combined with Artificial Bee Colony (ABC), micro-Artificial Bee Colony (μABC), Differential Evolution (DE), and Particle Swarm Optimization (PSO), in order to solve the SVM learning problem. The EA algorithms combined with KA were chosen based on good results shown solving different kinds of optimization problems and also for their exploration and exploitation capabilities, and low computational complexity [27,11,36,4,44,33,26].

Large-scale training algorithms for SVMs using EA is a promising field that has not been well explored. Although parallelization is a highly desirable approach to the large-scale classification problem, most large-scale SVM training algorithms do not take this into consideration to obtain better results in a shorter span of time. This is in part because testing complex parallel applications to guarantee a correct behavior is challenging. In scenarios such as where inherent data dependencies exist, a complex task cannot be partitioned because of sequential constraints, making parallelization less convenient [9,30]. One of the main goals in parallelizing an EA is to reduce the search time. This is a very important aspect for some classes of problems with firm search time requirements, such as in dynamic optimization problems and real-time planning [22].

2.4. THE KERNEL ADATRON ALGORITHM

The Adaptive Perceptron Algorithm (or Adatron) was first introduced by Anlauf and Biehl [3] in 1989 for linear classifier. This algorithm was proposed as a method for calculating the largest margin classifier. The Adatron is used for on-line learning Perceptrons and guarantees convergence to an optimal solution, when this exists [34].

In 1998, Fries et al. [17] proposed the KA algorithm. Basically, KA algorithm is an adaptation of Adatron algorithm for classification with kernels in high-dimensional spaces. It combines the simplicity of implementation of Adatron with the capability of a Support Vector Machine working in nonlinear feature spaces to construct a large margin hyperplane using on-line learning [32].

An advantage of using KA algorithm is the use of gradient ascent instead of quadratic programming, which is easier to implement and significantly faster to calculate.

To implement KA algorithm, it is necessary to calculate the dot product $w \cdot x_i$, where x_i is the set of training points and w denotes the normal vector to the hyperplane that divides the classes with a maximum margin. Since the kernel K is related to the high-dimensional mapping $\varphi(x_i)$ by the equation

$$K(x_i, x_j) = \varphi(x_i) \cdot \varphi(x_j), \qquad (2.3)$$

w can be expressed as

$$w = \sum_{i=1}^{n} \alpha i \gamma_i \varphi(x_i). \qquad (2.4)$$

Then, by using kernel K, the dot product can be expressed as

$$z_i = \sum_{j=1}^{d} \alpha_j \gamma_j K(x_i, x_j). \qquad (2.5)$$

To update the multipliers, a change in α_i must be proposed to be evaluated. The change can be calculated as follows:

$$\delta \alpha_i = \eta(1 - \gamma_i), \qquad (2.6)$$

$$\gamma_i = \gamma_i z_i, \qquad (2.7)$$

where η is the step size. If $\alpha_i + \delta \alpha i \leq 0$, then the proposed change to the multipliers would result in a negative α_i. To avoid this problem, set $\alpha_i = 0$. Otherwise, update $\alpha_i \leftarrow \alpha_i + \delta \alpha i$. The bias can be obtained as follows:

$$b = \frac{1}{2}(\min(z_i^+) + \max(z_i^-)), \qquad (2.8)$$

where z_i^+ are the patterns with class label $+1$, and z_i^- are those with class label -1.

The pseudocode is described briefly in Algorithm 6.

2.5. KERNEL ADATRON TRAINED WITH EVOLUTIONARY ALGORITHMS

As mentioned before, the KA algorithm requires the α_i value to be adjusted through iterations. In this approach, the adjustment is made using EA (Fig. 2.3). This type of algorithm was chosen as an optimization method because they are easy to implement, to parallelize, and have shown good results in diverse areas such as in computer vision, image processing, and path planning [27,11,4,6,12,2].

Algorithm 6 Kernel Adatron algorithm.

1: Initialize $\alpha_i = 1$.

2: **repeat**

3:     For (x_i, y_i) calculate z_i with Eq. (2.5).

4:     Calculate γ_i with Eq. (2.7).

5:     Calculate $\delta\alpha_i$ with Eq. (2.6).

6:     **if** $(\alpha_i + \delta\alpha_i) \le 0$ **then**

7:         $\alpha_i = 0$

8:     **end if**

9:     **if** $(\alpha_i + \delta\alpha_i) > 0$ **then**

10:         $\alpha_i = \alpha_i + \delta\alpha_i$

11:     **end if**

12:     Calculate b with Eq. (2.8)

13: **until** The stopping criteria are met.

The basic idea behind the proposed algorithms is to use a "divide and conquer" strategy, where each individual in the population of the EA (vector in DE, and bee in ABC and μABC) is seen as a subprocess; in this case, a thread that will solve a part of the whole problem. Once each subprocess has reached a result, it is compared to the results of its peers to improve future results.

DE, ABC, and μABC are easily parallelized due to the fact that each individual can be evaluated independently. The only phases where the algorithms require communication between their individuals are the phases that involve mutation and the selection of the fittest individual. Also the process to obtain the kernel matrix can be easily parallelized by dividing the process into several subtasks.

On each variant of the proposed algorithm, individual x_i (vector or bee) represents an n-dimensional vector composed of multipliers to be optimized over iterations by the EA.

The fitness function to be used by the algorithms is described by the equation

$$f(x) = \varepsilon * e^{abs(1-\Theta)}, \tag{2.9}$$

where ε is the number of elements misclassified, and Θ is the margin between classes of the hyperplane. The latter can be estimated as follows:

$$\Theta = \frac{1}{2}(\min(z_i^+) - \max(z_i^-)), \tag{2.10}$$

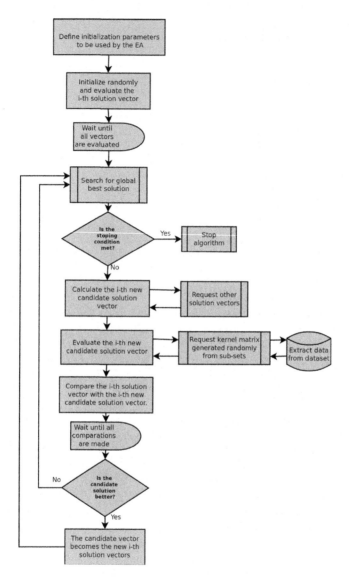

Figure 2.3 The diagram explains the basic idea behind the algorithm described in this chapter.

where z_i can be obtained by Eq. (2.5), z_i^+ are the patterns with class label $+1$, and z_i^- are those with class label -1.

The fitness function not only focuses on optimizing the margin of the hyper plane, but also it minimizes the classification error at the same time.

2.6. RESULTS USING BENCHMARK REPOSITORY DATASETS

A total of six data sets were used for the tests. They can be seen in Tables 2.1 and 2.2.

The Astro-ph dataset is focused on classifying abstracts of scientific papers from Physics ArXiv [24]. The aut-avn and real-sim classification datasets come from a collection of UseNet articles from four discussion groups: for simulated auto racing, simulated aviation, real autos, and real aviation. CCAT and C11 are datasets obtained from Reuters RCV1 collection, which poses the problem of separating corporate related articles [9]. The worm dataset focuses on classifying worm RNA splices [15].[1] The experiments were performed on an Intel Core i7-3770[2] machine with 16 GB of RAM and Fedora Linux 20[3] as operating system. The code was written in C++ using POSIX Threads[4] and Armadillo [37].

The bias term was disabled from the algorithms that appear in Table 2.2. As stopping condition, we used $f(x) \leq 0.1$ or 500 iterations.

For the EA fitness function, a linear kernel was used in all the algorithms since it gave us the best results in the generalization tests.

Previous to the tests, from each dataset a subset of 4000 training samples were extracted and normalized for binary classification and cross-validation. The results can be seen in Table 2.1. The generalization accuracy was obtained by applying a 10-fold cross-validation to each dataset. To test the accuracy of training capability of each algorithm, the SVM was trained using 400 training samples per run. The number of training samples was determined empirically, since it was observed that less than 400 training samples gave bad generalization results and more than 400 didn't improve the results considerably.

The values used to train the SVM with each EA were obtained empirically. For the μABC version of the algorithm, the following values were used: $RF = 0.85$, $C = 10$, and $FCR = 0.35$; the ABC version used: $C = 10$ and $lt = 75$; and the DE algorithm used: $C = 10$, $F = 0.85$ and $CROV = 0.25$. DE, and ABC versions used a swarm of 30 individuals each.

---

[1] The datasets can be found in http://www.csie.ntu.edu.tw/~cjlin/libsvmtools/datasets/binary.html and in http://users.cecs.anu.edu.au/~xzhang/data/.
[2] Intel and Intel Core are trademarks of Intel Corporation.
[3] Fedora is a trademark of Red Hat, Inc.
[4] For more information about POSIX: http://pubs.opengroup.org/onlinepubs/9699919799/.

Table 2.1 Density denotes the average percentage of nonzero elements of a data set

Data Set	Dimension	Density
Astro-ph	99757	0.08%
Aut-AVN	20707	0.23%
C11	47236	0.16%
CCAT	47236	0.16%
Real-Sim	20958	0.23%
Worm	804	25%

The main concern of implementing the algorithms described in this chapter is to work with the kernel matrix since its computational complexity is of $O(d * n^2)$, where d is the maximum number of nonzero features in any of the training samples, and n is the number of training samples. But, as can be appreciated in Table 2.1, the density of the data samples is really low in most of the cases. Therefore the number of operations to calculate the kernel matrix can be drastically reduced, and—if its treated as a divide and conquer problem the computational complexity is reduced—at worst case scenario, to $O(d * n^2/t)$, where t is the number of threads.

In most EAs, the computational complexity is due to fitness function evaluation. For the approaches shown in this chapter, the computational complexity of the fitness function is $O(n^2)$ and can be reduced to $O(n^2/t)$ by parallelization. A complexity of $O(n^2)$ is preferable to a problem of complexity of $O(d * n)$ for $d > n$, compared to SVM^{light} whose computational complexity is $O(d * n^2)$. The approaches shown in this chapter offer more appealing results, although for algorithms like OCAS and SVM^{perf}, computational complexity is $O(d * n)$ [30], which is preferable than the previously mentioned.

Though, as can be seen in Table 2.2, the approaches shown in this chapter cannot compete in terms of processing time and classification; the results shown in generalization tests are competitive with those shown by OCAS, SVM^{light}, and SVM^{perf}, (see Fig. 2.4). Something to be taken into account is that it is easier to implement and parallelize, EA algorithms rather than implement or parallelize the Quadratic Programming solvers used by OCAS [15], SVM^{light} [23], and SVM^{perf} [24].

Table 2.2 The generalization accuracy results (Gen.) were obtained by applying a 10-fold cross-validation to each dataset. The datasets were composed of 4,000 training samples. The training accuracy percentage and time were obtained by averaging the results of 15 tests on subsets of each dataset; each subset was composed of 400 training samples

Data Set	Alg.	Training	Gen.	Time
Astro-ph	μABC	93.2%	91.82%	0.669
	ABC	91.05%	88.55%	3.1505
	DE	92.52%	91.08%	0.3315
	SVM^{light}	100%	92.89%	0.020
	SVM^{perf}	100%	93.02%	0.061
	OCAS	99.75%	92.86%	0.109
Aut-avn	μABC	92.67%	89.84%	0.5275
	ABC	92.50%	89.51 %	1.668
	DE	92.52%	89.82%	0.054
	SVM^{light}	99.97%	90.95%	0.027
	SVM^{perf}	100%	93.02%	0.061
	OCAS	99.75%	90.93%	0.052
C11	μABC	75.47%	72.37%	0.832
	ABC	79.90%	75.35%	6.878
	DE	81.10%	76.43%	1.634
	SVM^{light}	99.67%	77.83%	0.021
	SVM^{perf}	99.80%	77.61%	0.065
	OCAS	99.65%	76.14%	0.075
CCAT	μABC	75.32%	73.05%	0.863
	ABC	80.70%	79.70%	6.333
	DE	82.55%	80.99%	1.110
	SVM^{light}	99.62%	82.76%	0.025
	SVM^{perf}	99.90%	82.78%	0.089
	OCAS	99.67%	83.17%	0.078
Real-Sim	μABC	93.90%	91.82%	0.256
	ABC	93.92%	91.90%	0.937
	DE	93.97%	92.06%	1.637
	SVM^{light}	99.87%	93.52%	0.018
	SVM^{perf}	99.87%	93.48%	0.062
	OCAS	99.57%	93.52%	0.036
Worm	μABC	90.00%	89.53%	0.717
	ABC	90.52%	89.91%	6.910
	DE	91.87%	90.65%	0.803
	SVM^{light}	99.17%	91.18%	0.027
	SVM^{perf}	100%	89.06%	0.028
	OCAS	99.75%	89.12%	0.145

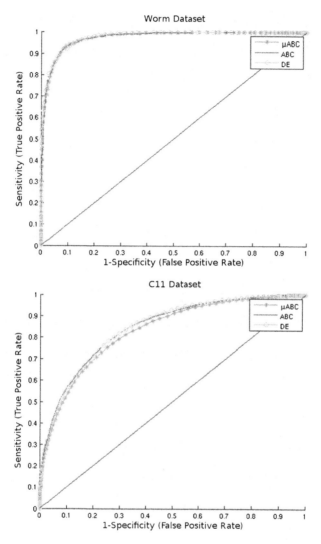

Figure 2.4 As can be seen from the ROC Curve, the generalization performance of the classifiers shown in this chapter is very similar.

 ## 2.7. APPLICATION TO CLASSIFY ELECTROMYOGRAPHIC SIGNALS

This section shows the application of the proposed SVM-EA system to classify electromyographic signals produced by movements of human upper limbs that represent hand motions; it is based on the work of [7].

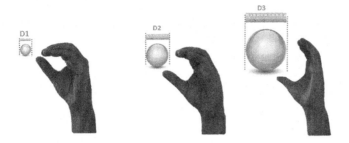

Figure 2.5 Grasping one of three different diameter spheres.

Electromyographic signals (sEMG) are a kind of myoelectric signals (MES), the electrical manifestation of the neuromuscular activation associated with a contracting muscle. They are time, and force, dependent whose amplitude varies randomly above and below the zero value [5]. The alteration in the temporal structure of these signals is high. However, statistics from waveform can be stable to allow pattern classification [21]. To date, it has not been possible to obtain a single parameter to appropriately reflect the characteristics of the MES measured. Because of this, a set of characteristics has to be chosen as input to a stage of pattern classification. Data for our experiments was obtained by placing surface electrodes on the forearm muscles of six subjects (three females F1, F2, and F3, and three males M1, M2, and M3). On each episode they were grasping one of three different diameter spheres (small D1, medium D2, and large D3) of expanded polystyrene (the weight is negligible); see Fig. 2.5. The subjects were seated comfortably, their arms were cleaned and scrubbed with alcohol and shaved to avoid high impedances in the signals. In each grasping episode, the sEMG of five forearm muscles were measured: flexor carpi radialis, flexor carpi ulnaris, flexor pollicis longus, flexor digitorum profundus, and extensor digitorum; see Fig. 2.6.

Each subject performed a batch, established as the grasp of three spheres (one at the time), with an interval of 5 seconds between grasping, and a lapse of time for relaxation between batches. Six contractions were contained in each batch (start to the end of a grasp), and the order in which the movements were performed was randomized for each subject. A brief contraction is performed at initial grasp of a sphere, indicating start of signal record. After that, subjects were instructed to maintain the grasp. Another contraction occurs after an audio cue, and the hand is opened, then returned to an anatomical rest position. The feature for extraction of Time Domain Features (TD) that are contained in MES in order to remove the

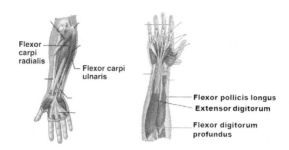

Figure 2.6 The sEMG was measured from these arm muscles.

Table 2.3 Training accuracy percentage results for datasets F1, F2, M1, and M2

Spheres $D_i - D_j$	Dataset F1 1-2 1-3 2-3	Dataset F2 1-2 1-3 2-3	Dataset M1 1-2 1-3 2-3	Dataset M2 1-2 1-3 2-3
OCAS	<u>0.97</u> 0.93 0.79	0.82 0.88 0.51	0.82 0.83 <u>0.74</u>	0.66 0.68 0.79
SVM^{perf}	<u>0.97</u> 0.96 0.90	<u>0.88</u> <u>0.95</u> <u>0.82</u>	0.81 0.83 0.72	0.67 0.72 0.80
KA-DE	<u>0.97</u> 0.93 <u>0.95</u>	0.85 0.90 0.66	<u>0.83</u> <u>0.84</u> 0.70	<u>0.74</u> <u>0.75</u> <u>0.83</u>
KA-PSO	<u>0.97</u> 0.92 0.93	0.85 0.90 0.60	<u>0.83</u> 0.83 0.69	0.71 0.73 0.81

unwanted EMG data was the one used in reference to the research developed by [21,38,45]. So the input data are the TD features of each sEMG of each subject on each grasping episode, and the class associated with this input data is represented for the size of the sphere that the subject is grasping during that episode. We had a total of three different classes. Each dataset was divided into four tests, two of them to evaluate the training capabilities of each classifier used, and the other two to evaluate generalization capabilities. During each test, the data was obtained from two grasping episodes. That is, one subject was grasping two spheres (one at a time), and then, we obtain three grasping episodes from each subject: grasping spheres D1–D2, D1–D3, and D2–D3; so each test is a two-class classification experiment. The datasets have a dimensionality of 40, and a total of 40 samples were used per test (20 samples per class), where 36 samples were used to train, and the rest was for testing the generalization capabilities of each classifier used. For generalization test, 10-fold cross-validation was used. Classifiers used were KA-PSO, KA-DE, OCAS, and SVM^{perf}. The results are shown in Tables 2.3 and 2.4.

Results demonstrated that it is possible to find a mapping between the pose of the fingers, and the object that they are grasping, despite the dis-

Table 2.4 Generalization accuracies results percentages

	Dataset F1	Dataset F2	Dataset M1	Dataset M2
Spheres $D_i - D_j$	1-2 1-3 2-3	1-2 1-3 2-3	1-2 1-3 2-3	1-2 1-3 2-3
OCAS	0.95 0.90 0.80	0.80 0.83 0.45	0.65 0.65 <u>0.53</u>	0.60 0.55 0.73
SVM^{perf}	0.95 <u>0.93</u> 0.88	<u>0.83</u> 0.88 <u>0.80</u>	0.68 <u>0.73</u> <u>0.53</u>	<u>0.68</u> 0.55 0.73
KA-DE	<u>0.98</u> 0.90 <u>0.95</u>	<u>0.83</u> 0.83 0.45	<u>0.75</u> <u>0.73</u> 0.50	0.65 <u>0.63</u> <u>0.75</u>
KA-PSO	0.97 0.90 0.94	<u>0.83</u> 0.81 0.44	<u>0.75</u> 0.71 0.50	0.65 0.61 0.73

turbances that can be found in the sEMG. The myoelectric datasets F1, F2, and M1 showed good results in the training phase. The best results were obtained using KA-DE followed closely by SVM^{perf}. For the generalization test, the results were not as good as the ones obtained on the training phase. Again, KA-DE and SVM^{perf} showed the best results, but it is worth mentioning that KA-DE is easier to implement, and parallelize, than SVM^{perf} and OCAS.

2.8. CONCLUSIONS

We developed a simple to implement method for classifying sparse large-scale datasets using parallelism with three EAs. The algorithms were designed to work principally with sparse data, but they can also work with not so sparse data. For example, the worm dataset gave good generalization results, and the times obtained are comparable with those obtained by the EA approaches in other tests.

The results of the tests show that the algorithms proposed in this chapter are competitive in terms of generalization, even though, OCAS, SVM^{light} and SVM^{perf} are superior in terms of execution time and the percentage of training accuracy. Comparing the three algorithms proposed, it is easy to notice that, in most of the tests, μABC version is faster, and DE version has the best accuracy.

Future improvements of the method will focus on these two versions, especially on the μABC approach. Such work will also address the reduction of computational complexity of the algorithm by reducing the number of training samples needed in each iteration, and the development of a multiclass version of this approach.

REFERENCES

[1] Adankon M, Cheriet M. Genetic algorithm-based training for semi-supervised SVM. Neural Comput Appl 2010;19(8):1197–206.

[2] AlRashidi M, El-Hawary M. A survey of particle swarm optimization applications in electric power systems. IEEE Trans Evol Comput 2009;13(4):913–8.

[3] Anlauf J, Biehl M. The Adatron: an adaptive perceptron algorithm. Europhys Lett 1989;10(7):687.

[4] Arana-Daniel N, Gallegos A, Lopez-Franco C, Alanis A. Smooth global and local path planning for mobile robot using particle swarm optimization, radial basis functions, splines and Bézier curves. In: 2014 IEEE congress on evolutionary computation (CEC). IEEE; 2014. p. 175–82.

[5] Basmajian J, Luca C. Muscles alive. Their functions revealed by electromyography. Williams and Wilkins; 1985.

[6] Benala T, Jampala S, Villa H, Konathala B. A novel approach to image edge enhancement using artificial bee colony optimization algorithm for hybridized smoothening filters. In: Nature & biologically inspired computing. IEEE; 2009. p. 1071–6.

[7] Benitez VH, Gallegos AA, Torres GA, Arana-Daniel N. Pattern recognition for finger position in structured scenarios. In: Proceedings of 21st Iberoamerican congress on pattern recognition. CIARP; 2016.

[8] Bordes A, Bottou L, Gallinari P. SGD-QN: careful quasi-Newton stochastic gradient descent. J Mach Learn Res 2009;10:1737–54.

[9] Bottou L, Chapelle O, DeCoste D, Weston J. Large-scale kernel machines. Neural Information Processing. MIT Press; 2007.

[10] Cortes C, Vapnik V. Support vector networks. Mach Learn 1995;20:273–97.

[11] Das S, Suganthan P. Differential evolution: a survey of the state-of-the-art. IEEE Trans Evol Comput 2011;15(1):4–31.

[12] de Fraga L. Self-calibration from planes using differential evolution. In: Progress in pattern recognition, image analysis, computer vision, and applications. Springer; 2009. p. 724–31.

[13] Fawcett T. "In vivo" spam filtering: a challenge problem for KDD. SIGKDD Explor Newsl Dec. 2003;5(2):140–8.

[14] Floudas C, Visweswaran V. Quadratic optimization, vol. 2. Springer US; 1995.

[15] Franc V, Sonnenburg S. Optimized cutting plane algorithm for large-scale risk minimization. J Mach Learn Res 2009;10:2157–92.

[16] Frasch J, Sager S, Diehl M. A parallel quadratic programming method for dynamic optimization problems. Math Program Comput 2013:1–41.

[17] Frieß T, Cristianini N, Campbell C. The Kernel-Adatron algorithm: a fast and simple learning procedure for support vector machines. In: Proceedings of the fifteenth international conference on machine learning. Morgan Kaufmann; 1998.

[18] Garcia E, Rangel P, Lozano F. Adaptive support vector machines for time series prediction. In: Proceedings of XI simposio de tratamiento de señales, imagenes y vision artificial; Sep. 2006.

[19] Haykin S. Neural networks: a comprehensive foundation. 3rd ed. Prentice-Hall; 2007.

[20] Huaitie X, Guoyu F, Zhiyong S, Jianjun C. Hybrid optimization method for parameter selection of support vector machine. In: IEEE international conference on intelligent computing and intelligent systems (ICIS), vol. 1; Oct. 2010. p. 613–6.

[21] Hudgins B, Parker P, Scott R. A new strategy for multifunction myoelectric control. IEEE Trans Biomed Eng 1993;40:82–94.

[22] Indiveri G. Handbook of computational intelligence. Berlin, Heidelberg: Springer-Verlag; 2015.

[23] Joachims T. Making large scale SVM learning practical. In: Advances in kernel methods – support vector learning. MIT Press; 1999.

[24] Joachims T. Training linear SVMs in linear time. In: Proceedings of the conference on knowledge discovery and data mining; 2006.

[25] Kao J, Chuang J, Wang T. Chromosome classification based on the band profile similarity along approximate medial axis. Pattern Recognit 2008;41(1):77–89.

[26] Karaboga D, Akay B, Ozturk C. Artificial bee colony (ABC) optimization algorithm for training feed-forward neural networks. In: Modeling decisions for artificial intelligence. Springer; 2007. p. 318–29.

[27] Karaboga D, Gorkemli B, Ozturk C, Karaboga N. A comprehensive survey: artificial bee colony (ABC) algorithm and applications. Artif Intell Rev 2014;42(1):21–57.

[28] Kivinen J, Smola A, Williamson R. Online learning with kernels. IEEE Trans Signal Process 2004;52(8):2165–76.

[29] Lessmann S, Stahlbock R, Crone S. Genetic algorithms for support vector machine model selection. In: International joint conference on neural networks; 2006. p. 3063–9.

[30] Menon A. Large-scale support vector machines: algorithms and theory. Research exam. San Diego: University of California; 2009. p. 1–17.

[31] Mierswa I. Evolutionary learning with kernels: a generic solution for large margin problems. In: Proceedings of the 8th annual conference on genetic and evolutionary computation (GECCO '06). New York, NY, USA: ACM Press; 2006. p. 1553–60.

[32] Müller K, Mika S, Rätsch G, Tsuda K, Schölkopf B. An introduction to kernel-based learning algorithms. IEEE Trans Neural Netw 2001;12(2):181–201.

[33] Noman N, Iba H. Differential evolution for economic load dispatch problems. Electr Power Syst Res 2008;78(8):1322–31.

[34] Opper M. Learning times of neural networks: exact solution for a perceptron algorithm. Phys Rev A 1988;38(7):3824.

[35] Osowski S, Siwek K, Markiewicz T. MLP and SVM networks: a comparative study. In: Proceedings of the 6th Nordic signal processing symposium; 2004. p. 9–11.

[36] Rajasekhar A, Das S, Das S. μABC: a micro artificial bee colony algorithm for large-scale global optimization. In: GECCO'12 companion. ACM; 2012. p. 1399–400.

[37] Sanderson C. Armadillo: an open source C++ linear algebra library for fast prototyping and computationally intensive experiments. Tech. rep. NICTA; 2010.

[38] Sapsanis C, Georgoulas G, Tzes A, Lymberopoulos D. Improving EMG based classification of basic hand movements using EMD. In: Proceedings of engineering in medicine and biology society, EMBC, 35th annual international conference of the IEEE; 2013. p. 5754–7.

[39] Shalev-Shwartz S, Nathan S. SVM optimization: inverse dependence on training set size. In: Proceedings of the 25th international conference on machine learning; 2008. p. 928–35.

[40] Shalev-Shwartz S, Singer Y, Nathan S. Pegasos: primal estimated sub-gradient solver for SVM. In: Proceedings of the 24th international conference on machine learning; 2007. p. 807–14.

[41] Sudholt D. Parallel evolutionary algorithms. In: Handbook of computational intelligence. Springer; 2015. p. 929–59.

[42] Teo C, Smola A, Vishwanathan S, Le Q. A scalable modular convex solver for regularized risk minimization. In: Proceedings of the 13th ACM SIGKDD international conference on knowledge discovery and data mining; 2007. p. 727–36.

[43] Vavasis S. Complexity theory: quadratic programming. In: Encyclopedia of optimization. Springer US; 2009. p. 451–4.

[44] Yoshida H, Kawata K, Fukuyama Y, Takayama S, Nakanishi Y. A particle swarm optimization for reactive power and voltage control considering voltage security assessment. IEEE Trans Power Syst 2000;15(4):1232–9.

[45] Zecca M, Micera S, Carrozza M, Dario P. Control of multifunctional prosthetic hands by processing the electromyographic signal. Crit Rev Biomed Eng 2002;30.

CHAPTER THREE

Reconstruction of 3D Surfaces Using RBF Adjusted with PSO

Contents

3.1. INTRODUCTION

This chapter presents the application of Radial Basis Functions Neural (RBF) networks trained with PSO to solve the problem of obtaining surfaces from point-clouds derived from 3D range scanners.

Meshing functions in order to interpolate point-clouds to obtain compact surface representations is very important to Computer Aided Design (CAD), robot mapping, and object description and recognition among other important applications. Nevertheless, the quantifiers about the precision, and quality, of the interpolation are different between applications. In CAD, it is very important to obtain smooth surfaces and to ensure that they are manifolds. Meanwhile, for robot mapping and object description and recognition, it is essential to get compact (low memory and dimension) and rich-in-information or representative-enough representations of the objects or environments. That is to say, it is important that the obtained surface describes the shape of the object or the environment without having great detail (or without ensuring the manifold condition). In this paper, we propose an approach having to do with using RBFs trained with PSO to fulfill all the precision and quality requirements for all the applications. That is, our approach can be configured to obtain compact representation, and not great-detailed surfaces, to be used for robot mapping or object recognition, as well as to obtain smooth blending surfaces and ensuring that they are manifold. RBFs have been used before to reconstruct surfaces

© 2018 Elsevier Inc.
All rights reserved.
33

from point-clouds [1] due to their compact, functional description, and that the interpolation and extrapolation are inherent in their functional representation. Nevertheless, they have been trained using gradient-descendent methods, which are very susceptible to getting trapped into local minima, and also they are not-so-easy to be parallelized. We propose a training methodology for RBFs based on PSO, and our obtaining an interpolating system for point-clouds that is precise and fast due to its parallelization. The advantages of getting surfaces with RBFs have been reported before by Savchenko [4], Turk and O'Brien [6], and Carr [2]. However, these works had not been proposed to deal with large data sets, as the ones obtained with LIDAR sensors. The system presented in [1] proposed great advantages to this issue but using classic training methodologies for RBFs that are computationally expensive and slow to converge.

Next, we present our solution to this issue, based mainly on using the PSO algorithm as training methodology for the RBFs to get a parallelized algorithm that is faster to converge, and yet, precise enough to produce smooth and representative surfaces of dense point-clouds.

3.2. RADIAL BASIS FUNCTIONS

This kind of neural networks has an architecture with three layers: an input layer, one hidden layer, and an output layer as represented in Fig. 3.1.

Although the architecture of the network can be similar to the architecture of a Multilayer Perceptron (MLP), the main difference is that the hidden neurons compute the Euclidean distance between the weight vector (which, in RBF networks, is named as center or centroid) and the input vector. This computed distance is the evaluation parameter of a radial basis function with Gaussian shape.

There are several methods to train the hidden layer of an RBF; the most used and known is the algorithm named K-means. That is the one we used in this chapter. K-means is a clustering algorithm that uses a nonsupervised type of learning, where K is the number of clusters that we wish to find, and this is the number of the hidden layer. The latter is a user parameter, and it has to be fixed. The steps of the algorithm K-means are as follows:

1. Initialize the weight vector (the centers or centroids of each cluster). Typically initializing the weight vector is made by assigning the values of the first k samples of the training data set: $c_1 = x_1, c_2 = x_2, \ldots, c_k = x_k$, where c_j is the jth center (or weight vector), x_i is the ith sample of

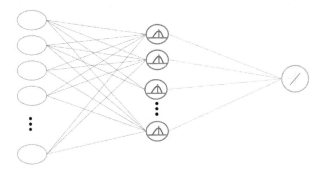

Figure 3.1 The three-layer architecture of an RBF. The first layer is the input layer. The hidden layer is the radial basis units layer, and the output layer is a linear unit layer.

the training data set, where $j = 1, \ldots, k$ and $i = 1, \ldots, N$; N is the total number of training samples.

2. During each iteration, the samples of the data are assigned to a cluster: For a sample data x_i, the distance between x_i and each one of the cluster centers c_j is computed. The sample x_i is assigned to the cluster with a shorter distance to the center.

3. The new center vectors c_j are computed for each cluster. They are computed as the mean vector of the sample vectors which belong to each cluster. The above is the same as computing the mass center of each pattern distribution, considering that all the patterns in the cluster weight the same.

4. If the values of the centers change with respect to the previous iteration, then go to step 2. Otherwise, the convergence is reached, and the training process is ended.

The K-means algorithm ensures the convexity and also divides the input set of data in a partition, named the Voronoi partition as it is shown in Fig. 3.2.

Once the centers are fixed, the standard deviation of each Gaussian function has to be computed. The standard deviation of each neuron is computed as in Eq. (3.1) for all x in the cluster.

$$\sigma = \sqrt{E\left[(X - c_k)^2\right]} \tag{3.1}$$

Herein, E denotes the average or expected value, and c_k is the mean vector value of each cluster.

Then, the centers and spreads of each Gaussian function are now fixed as is shown in Fig. 3.3.

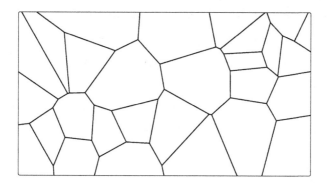

Figure 3.2 Voronoi partition.

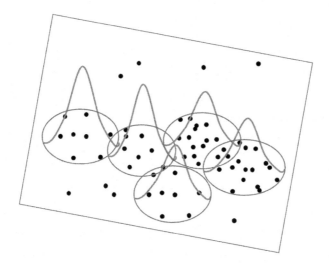

Figure 3.3 Centers of Gaussian basis functions on the input space.

Finally, the output weights are adjusted to interpolate a function $y : \Re^n \to \Re$. As the output of the network is a scalar function of the input vector, this network output is given by the equation

$$F(x) = \sum_{i=1}^{N} w_i \varphi(||x - c_i||), \qquad (3.2)$$

where $\varphi(||x - c_i||) = exp^{\frac{-(||x-c_i||)}{2\sigma^2}}$, and the norm is typically the Euclidean or the Mahalanobis distance.

So, the weights values can be obtained by solving the following equation:

$$
\begin{bmatrix}
\varphi_1(x_1) & \varphi_2(x_1) & \varphi_3(x_1) & \cdots & \varphi_N(x_1) \\
\varphi_1(x_2) & \varphi_2(x_2) & \varphi_3(x_2) & \cdots & \varphi_N(x_2) \\
& \vdots & & & \\
\varphi_1(x_N) & \varphi_2(x_N) & \varphi_3(x_N) & \cdots & \varphi_N(x_N)
\end{bmatrix}
\begin{bmatrix}
w_1 \\ w_2 \\ \vdots \\ w_N
\end{bmatrix}
=
\begin{bmatrix}
y_1 \\ y_2 \\ \vdots \\ y_N
\end{bmatrix}, \tag{3.3}
$$

where the interpolation matrix with elements $[\varphi_{iq}] = \Phi$ for $i, q = 1, \ldots, N$. Defining the vectors $\mathbf{w} = [w_i]$ and $\mathbf{y} = [y_i]$, Eq. (3.3) can be written as

$$\Phi\mathbf{w} = \mathbf{y}. \tag{3.4}$$

By the theorem of Michelli (1986), if the points x_i are distinct, then the interpolation matrix Φ is nonsingular, and the weights $\mathbf{w}$ can be solved using simple linear algebra:

$$\mathbf{w} = \Phi^{-1}\mathbf{y}. \tag{3.5}$$

3.3. INTERPOLATION OF SURFACES WITH RBF AND PSO

In this section, the proposal of training RBFs using PSO is presented. Once the RBF is trained, the 3D surfaces are drawn using the ellipsoid of covariance to set the boundaries of the radial basis functions that define the 3D surface.

So, as the architecture of RBF and its classic training algorithm have been presented in the above section, and the PSO algorithm is shown in Chapter 1 of this book. In the next subsection the method to compute the ellipsoid of covariance is introduced.

3.3.1 Ellipsoid of covariance

The ellipsoid of covariance is a geometric representation of the data distribution. If we have a set of points $\mathbf{x}$ and their 3D coordinates (x_i, y_i, z_i) with $i = 1, 2, \ldots, n$, then we can define the covariance as

$$\sigma(x, y) = \frac{1}{n}\sum_{i=1}^{n}(x_i - \mu_x)(y_i - \mu_i). \tag{3.6}$$

Therefore, the covariance matrix (Eq. (3.8)) has all the variances and covariances of the data $\mathbf{x}$. To calculate the ellipsoid of covariance, it is

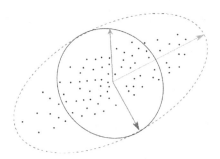

Figure 3.4 Covariance ellipsoid.

necessary to compute the eigenvalues and eigenvectors of this matrix. But first, the covariance matrix has to be scaled using an inverse-χ-squared-distribution. The above follow suit because the sums of squares of Gaussian random variables follow a chi-square distribution. A chi-square distribution is defined in terms of "degrees of freedom", which represent the number of unknowns. In our case, there are three unknowns (three semiaxes of the ellipsoid), and therefore three degrees of freedom as it is shown is Eq. (3.7) of the 3D ellipse.

$$\left(\frac{x}{\sigma_x}\right)^2 + \left(\frac{y}{\sigma_y}\right)^2 + \left(\frac{z}{\sigma_z}\right)^2 = s. \tag{3.7}$$

So the covariance ellipsoid is scaled by a value s that follows a chi-square distribution, and we have to "eliminate" this scale factor by dividing the elements of the matrix by the inverse of the value s. We use a value of $s = 0.0.584$ that represents 90.

Finally, the eigenvectors represent the unit vectors of the semiaxes of an ellipsoid. If we scale these with the root of the eigenvalues, we can get an ellipsoid as shown in Fig. 3.4.

$$\Sigma = \begin{bmatrix} \sigma(x,x) & \sigma(x,y) & \sigma(x,z) \\ \sigma(y,x) & \sigma(y,y) & \sigma(y,z) \\ \sigma(z,x) & \sigma(z,y) & \sigma(z,z) \end{bmatrix}. \tag{3.8}$$

We can use the scaled covariance matrix Eq. (3.8) to calculate the Mahalanobis distance (Eq. (3.9)). The computation of this distance has lower computational cost than the Euclidean distance and allows us to decide if one point falls inside $(D_M(\mathbf{x}) < 1)$ or outside $(D_M(\mathbf{x}) > 1)$ the ellipsoid.

$$D_M(\vec{x}) = \sqrt{(\vec{x} - \vec{\mu})^T \Sigma^{-1} (\vec{x} - \vec{\mu})}. \tag{3.9}$$

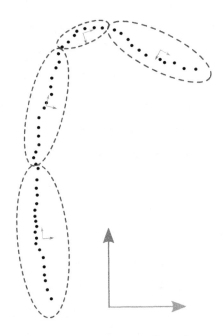

Figure 3.5 Mapping scheme using 2D covariance ellipsoids.

The covariance ellipsoid is used in several applications of Artificial Intelligence. In fuzzy, logic is used to model membership functions [3]. The covariance ellipsoid is useful also to solve mapping tasks. For example, if K-means is used as segmentation algorithm of a point cloud, then the obtained clusters can be represented by an ellipsoid as it is shown in Fig. 3.5.

Using a benchmark point-cloud as it is the Stanford Bunny model [5] shown in Fig. 3.6 and obtaining clusters of its point-cloud using ellipsoids of covariance. The result is shown in Fig. 3.7.

3.3.2 RBF-PSO and ellipsoid of covariance to interpolate 3D point-clouds

In this subsection, we present our algorithm to solve the problem of obtaining the 3D surface of a point-cloud. Fig. 3.8 shows a simplified scheme of our proposal. The steps of the process are described as follows:

1. Point-cloud segmentation. The point-cloud is segmented using the ellipse of covariance in a supervised manner into clusters S_i. The supervised phase is done to ensure that the portion of the surface that the cluster contains can be approximated using a linear combination of functions by the RBF. That is to say, each ellipse of covariance

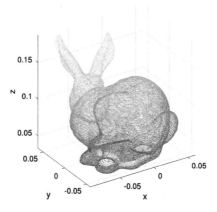

Figure 3.6 Stanford Bunny Point Cloud. The point cloud has a total size of 362,272 points obtained with a Cyberware 3030 MS scanner.

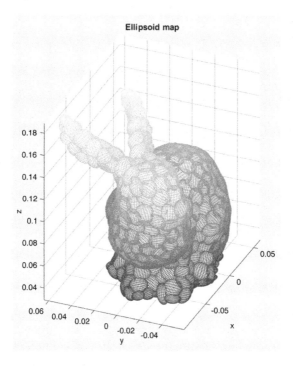

Figure 3.7 Ellipsoids of covariance on each cluster obtained using K-means.

and the portion of points that it contains are tested to ensure that the surface described by them could be approximated using a function surface.

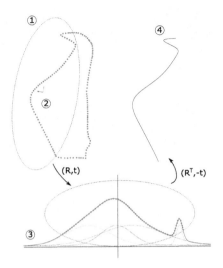

Figure 3.8 Simplified scheme of the proposed algorithm.

2. Rigid transformation computation. A rigid transformation $T = (R, t)$ (rotation and translation transformation) is computed in order to move the portion of the surface segmented into a common coordinate system.

3. The RBF is trained using K-means and PSO. The centers μ_x, μ_y and standard deviations σ_x, σ_y of each RBF are determined using K-means. Afterward, using a number of particles that is equal to the number of functions that we want to use for cluster and the fitness function in Eq. (3.10), the RBF is trained to obtain an interpolation of the point-cloud surface. Meaning, we adapt the z-coordinate of each point (x_i, y_i, z_i) in the set of training (in the input point-cloud).

4. Finally, the surface obtained is drawn. Using the inverse of the rigid transformation computed in step 1, $T^{-1} = (R^T, -t)$, and the Mahalanobis distance (Eq. (3.9)), the obtained interpolation surface is drawn.

Formally, in Fig. 3.9, the flowchart of the proposed algorithm described above is shown.

The fitness function for PSO that we designed has as objective for each point in the segment of the point cloud (x_i, y_i, z_i) to minimize the square root of mean square error between the z-coordinate of the transformed point cloud and the estimated output of the RBF for the input (x, y) as can

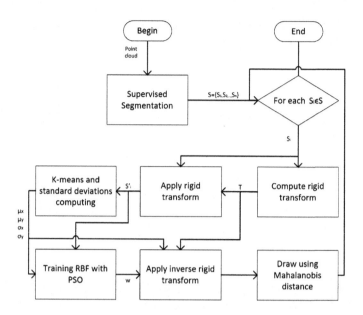

Figure 3.9 Flowchart of the proposed algorithm.

be seen in the equation

$$f = \min_{w_1, w_2, \ldots, w_k} \sqrt{\sum_{\bar{x} \in S_i} \frac{\left(z - \sum_{j=1}^{k} w_j e^{-\left[\frac{(x-\mu_x)^2}{2\sigma_x^2} + \frac{(y-\mu_y)^2}{2\sigma_y^2} \right]} \right)^2}{|S_i|}}. \tag{3.10}$$

3.3.3 Experimental results

In this subsection, the experimental results are shown. Firstly, experimental results using RBF trained with PSO for 2D benchmark function approximation are presented.

These experiments were conducted in order to prove the effectiveness of the RBF-PSO training methodology before we designed the method shown in Figs. 3.8 and 3.9 for 3D shapes. So, the fitness function that we used is a simplified version of Eq. (3.10), given the dimension reduction. In these experiments, for 2D function approximation, the objective is for each point in the function (that is sampled with N points) surface to minimize the square root of the mean square error between the y-coordinate of the point on the function, and the estimated output of the RBF for the input x. This can be seen in the equation

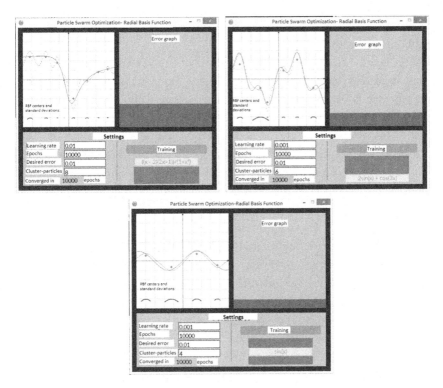

Figure 3.10 PSO-RBF results for approximating 2D functions. (For interpretation of the references to color in this figure, the reader is referred to the web version of this chapter.)

$$f = \min_{w_1, w_2, \ldots, w_k} \sqrt{\sum_{i=1}^{N} \frac{\left(y - \sum_{j=1}^{k} w_j e^{-\left[\frac{(x-\mu_x)^2}{2\sigma_x^2} \right]} \right)^2}{|N|}}. \tag{3.11}$$

In Fig. 3.10, there are shown the results for approximating three 2D benchmark functions. The function surface is presented as a solid black line. Meanwhile, the approximation result is shown as a dotted green line. The RBF centers and standard deviations are shown in the bottom of the 2D plane using blue lines. The number and final positions of particles correspond to red crosses. Settings of each experiment are shown on the bottom of each figure.

As the reader can note, using few particles in each experiment (eight, six, and four), the results are precise enough for each function. These simple, but promising, results motivated us to design the method in Fig. 3.9 to interpolate 3D surfaces.

Next, in Fig. 3.11 are shown different views of the 3D surface obtained using our algorithm and the point-cloud of the Stanford bunny as input data. An approximation with 117 clusters was used with three particles per cluster, each one representing an RBF unit. It is important to mention that in step 1 of our process, in which the supervised segmentation using the ellipse of covariance is performed, some clusters are discarded because their surfaces did not meet the requirements to be a function. Therefore, it would be impossible to compute an interpolation using RBFs. The foregoing is why, in the figure of results, some holes can be seen in the surface. But on the other hand, our process of approximating the surface has lower computational complexity than the classic algorithm for training RBFs because the latter implies computing an inverse of a matrix. Meanwhile, the PSO is about computing two linear equations per particle, and each particle can be seen as a subprocess; in this case, a thread that will solve a part of the whole problem. Once each subprocess has reached a result, it is compared to the results of its peers to improve future results.

PSO is easily parallelized due to the fact that each particle can be evaluated independently. The only phases where the algorithms require communication between their individuals are the phases that involve mutation, and the selection of the fittest individual.

The error of each RBF unit used in SBSA experiment is shown in Fig. 3.12. As can be seen, the first clusters have the largest error of approximation. This is because these clusters have more points than the last clusters in the graph (clusters approximated with RBF units number 70 and up in Fig. 3.12), and the lower number of points in the cluster, the smaller approximation error obtained.

Another experiment result was obtained using as input the point-cloud provided by an SRI-500 Laser Rangfinder from Acuity Technologies capable of scanning distances up to 500 feet and 800,000 points per second. The number of points in the cloud shown in Fig. 3.13 is 794,917. The clusters obtained with K-means are shown in the upper side of Fig. 3.14. In this experiment, no cluster was eliminated because it did not fulfill the condition of being a surface of a function. This was due to the clusters having a small standard deviation, and therefore the result of RBF-PSO approximation has no holes in the surface as shown in the bottom side of Fig. 3.14.

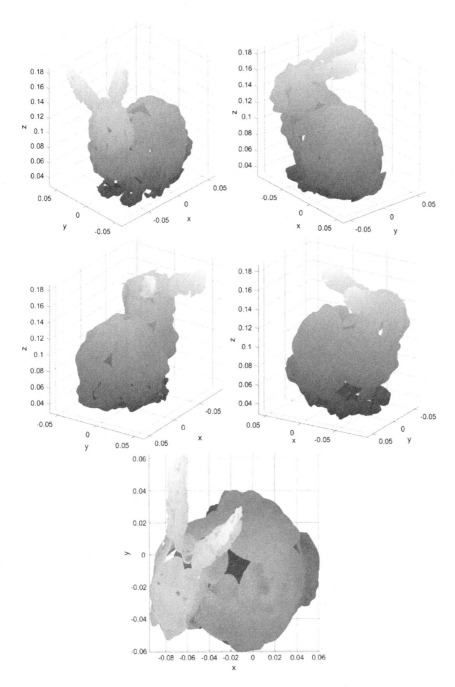

Figure 3.11 Results of RBF-PSO surface approximation using Stanford Bunny as input data.

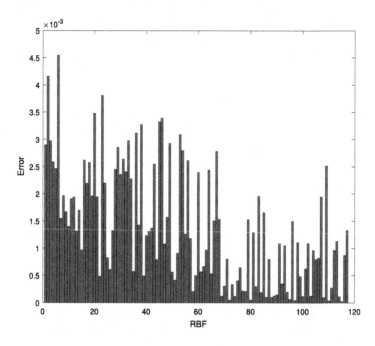

Figure 3.12 Final error per RBF unit. To approximate the SBS, 117 RBF units were used.

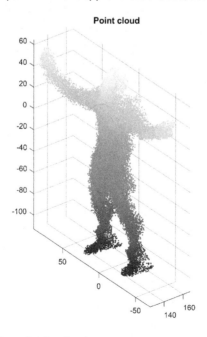

Figure 3.13 Human-shaped point cloud.

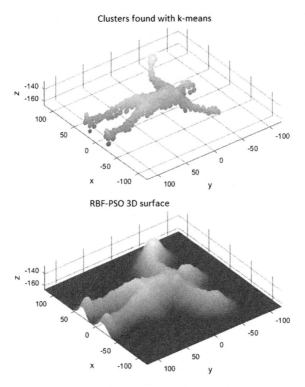

Figure 3.14 Upper side: Clusters obtained with *K*-means algorithm. Bottom side: RBF-PSO surface obtained.

3.4. CONCLUSION

In this chapter, an algorithm to approximate non-structured shapes using Radial Basis Functions networks adapted with Particle Swarm Optimization was proposed. The ellipse of covariance was computed to set the boundaries of the radial basis functions that define the 3D surface, as well as to allow us to decide if one point falls inside or outside the ellipsoid. The results prove that our proposal is useful to obtain 3D surfaces that describe the shapes of the objects, and although the algorithm does not produce rendered shapes, the obtained surfaces can be used as descriptors for pattern recognition process and environmental mapping. The reason for the same being that it is fast enough to be implemented in real-time, which is necessary for robot navigation tasks. On the other hand, the reduction of the number of parameters used to describe a shape with 3D point clouds (number of points in the cloud × 3) is significant. For example, the Stanford bunny can be described by 117 parameters of RBFs (μ_x, μ_y, σ_x, σ_y, and z-coordinate) instead of 362,272 × 3 parameters of the point cloud. In the future, we will work on an autonomous algorithm to segment the cloud point and test if the portion of the surface contained in the segments meets the surface of a function to replace the supervised part of our process. This could avoid having holes in the final obtained surface, and facilitate our getting rendered surfaces of the objects.

REFERENCES

[1] Carr JC, Beatson RK, Cherrie JB, Mitchell TJ, Fright WR, McCallum BC, et al. Reconstruction and representation of 3D objects with radial basis functions. In: Proceedings of the 28th annual conference on computer graphics and interactive techniques, ser. SIGGRAPH '01. New York, NY, USA: ACM; 2001. p. 67–76. [Online]. Available from: http://doi.acm.org/10.1145/383259.383266.

[2] Carr J, Beatson R, Fright W. Surface interpolation with radial basis functions for medical imaging. IEEE Transactions on Medical Imaging Feb 1997;16(1).

[3] Dickerson JA, Kosko B. Fuzzy function learning with covariance ellipsoids. In: Neural networks, 1993, IEEE international conference on. IEEE; 1993. p. 1162–7.

[4] Savchenko VV, Pasko AA, Okunev OG, Kunii TL. Function representation of solids reconstructed from scattered surface points and contours. Computer Graphics Forum 1995;14(4):181–8.

[5] Turk G, Levoy M. The Stanford bunny. Wolfram Research; 2016. https://doi.org/10.24097/wolfram.91305.data.

[6] Turk G, O'Brien JF. Shape transformation using variational implicit functions. In: Proceedings of the 26th annual conference on computer graphics and interactive techniques, ser. SIGGRAPH '99. New York, NY, USA: ACM Press/Addison-Wesley Publishing Co.; 1999. p. 335–42. [Online]. Available from: https://doi.org/10.1145/311535.311580.

CHAPTER FOUR

Soft Computing Applications in Robot Vision

Contents

4.1. INTRODUCTION

Robots require the use of sensors to measure its environment, to achieve different tasks, like grasping, obstacle avoidance, path planning, robot navigation, and 3D map reconstruction. There are many sensors used by robots. However, one of the most used types is vision sensors. The advantages of vision sensors are their low cost, low power consumption, and the great information that they can provide. Traditionally, the most common vision sensors are the perspective cameras. Recently, RGB-D cameras became popular due to their cost and the data they produce. The sensors they carry provide RGB images along with per-pixel depth information.

In recent years, soft computing algorithms have gained much popularity to solve engineering applications. For that reason, we were interested in the application of soft computing algorithms to solve important computer vision problems like image tracking and plane detection. These capacities are very important since they can be used to perform more complex tasks, like grasping and robot navigation.

Bio-inspired Algorithms for Engineering
https://doi.org/10.1016/B978-0-12-813788-8.00004-4
© 2018 Elsevier Inc.
All rights reserved.
49

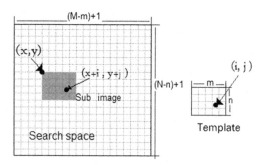

Figure 4.1 Template matching.

In the following sections, we will describe the image tracking and plane detection problems, and present the proposed approach using a soft computing algorithm.

4.2. IMAGE TRACKING

The technique used for image tracking in this work is the template matching approach. Template matching tracking is a subclass of kernel tracking. The objective of this method is to determine the location of a desired pattern represented by a template image inside an image. The technique is also used to measure the similarity between a target image and a template image.

During template tracking, the template is shifted pixel by pixel across the image, computing correlation plane that provides information of where the template best matches the image (Fig. 4.1).

In this section, we present a method for image tracking based on PSO [1], that is, in this method, PSO particles are used to find the best template match.

4.2.1 Normalized cross correlation

Let $I(m, n)$ represent the intensity value of the image, and let T represent the template image. We assume that the size of the image I is $M \times N$ and the size of the template is $m \times n$. Obviously, we assume that the size of I is greater than the size of T. Our goal was to find the best position of the template T in the image I. To measure the similarity between the patch image and the template, we used the normalized cross-correlation (NCC).

The Normalized Cross Correlation (NCC) is an effective method to measure similarity, which is invariant to linear brightness and contrast variations. The NCC between the image I and the template T is defined as

$$NCC(u, v) = \frac{\sum_{x,y}[I(x, y) - \bar{I}_{u,v}][T(x - u, y - v) - \bar{T}]}{\sqrt{\sum_{x,y}[I(x, y) - \bar{I}_{u,v}]^2 \sum_{x,y}[T(x - u, y - v) - \bar{T}]^2}}, \quad (4.1)$$

where $\bar{I}_{u,v}$ is the mean of the image region centered at u, v, and $\bar{T}$ represents the mean of the template.

The values of $NCC(u, v)$ are defined in the range of $[-1.0, 1.0]$, where a value of 1.0 means a perfect match. The template in the image is located by searching for the maximum of the NCC function.

The tracking algorithm involves sliding the template over the search area. At each position, the algorithm computes the NCC, which measures the degree of similarity between the template and the image. The algorithm stops when it finds the best match. More information about NCC can be found in [2].

4.2.2 Continuous plane to image plane conversion

As we know, the PSO algorithm is designed for continuous spaces. Since the images are discrete, we propose to work on a continuous normalized plane and then convert the results into coordinates on an image plane. We considered the following equations for the conversion:

$$\rho_x = round\left(\frac{W}{2}(\alpha_x + 1)\right), \quad (4.2)$$

$$\rho_y = round\left(\frac{H}{2}(1 - \alpha_y)\right), \quad (4.3)$$

where α_x and α_y are the positions on the normalized plane, ρ_x and ρ_y are the positions on the image plane, and W and H are the width and height of the image, respectively.

Fig. 4.2 shows that each individual position (α_x, α_y) in the normalized plane corresponds to an image plane position (ρ_x, ρ_y).

4.2.3 Algorithm implementation

The algorithm begins with the initialization of the PSO particles with random values from -1 to 1. Therefore, all the particles are defined inside the

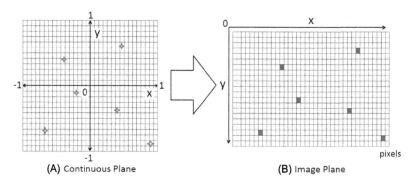

(A) Continuous Plane   **(B)** Image Plane

Figure 4.2 Conversion from continuous plane to image plane: (A) PSO particles, (B) corresponding particles in the image plane.

boundaries of the continuous plane (Fig. 4.2). Then, each particle position is converted to pixels in an image plane.

The PSO particles move inside the plane and search the position with the best fitness value, which represents the perfect match. The fitness function consists in the evaluation of the NCC measure.

The objective function can be defined using (4.1) as follows:

$$f(\rho_x, \rho_y) = -NCC(\rho_x, \rho_y). \tag{4.4}$$

The best solution represents the position (ρ_x, ρ_y) of the template T located into the image I. This solution is found by solving a minimization problem stated as

$$\min(f(\rho_x, \rho_y)) \text{ subject to } \vec{l}_b \le \vec{x}_{i,G} \le \vec{u}_b.$$

A summary of the proposed method is shown in Algorithm 7. The stop criterion is met when the proposed algorithm reaches the total number of iterations.

In order to reduce computational time, we used the image pyramid method. Therefore, our algorithm computes the search in the coarse image and then refines the search in finer images.

4.2.4 Experiments

The experiments were performed using a Kinect sensor, an RGB-D sensor. The RGB-D sensor is able to provide an RGB image with depth information related to it.

Algorithm 7 Algorithm of the proposed approach.

1: Randomly initialize particles with values from -1 to 1
2: **repeat**
3:     Convert particles position to pixels position using (4.2) and (4.3)
4:     Evaluate the fitness function (4.4) for each particle $\vec{x}_i$
5:     Determine global best γ
6:     **for all** particles **do**
7:         Update particle velocity $\vec{v}_i$
8:         Update particle position $\vec{x}_i$
9:     **end for**
10: **until** Stop criterion is met

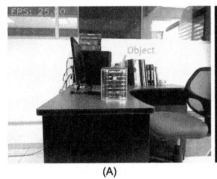 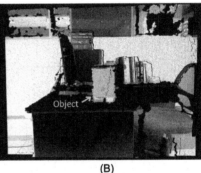

(A)   (B)

Figure 4.3 (A) Template matching results on the image, (B) 3D results using the depth data.

The algorithm performs its search in the RGB image using the PSO algorithm to detect the best match. Once the algorithm detects the best match, we were able to use the depth information of the object. Thus, the algorithm proved its ability to track the object in the 3D space, as indicated in Fig. 4.3.

4.3. PLANE DETECTION

One important task in robotics is the detection of obstacles. In the case of indoors environments the most common geometric object is the plane. In this section, we present an approach to estimate the planes in a scene.

The most common sensors to detect planes are cameras, and lasers. However, the RGB-D sensors have gained the attention of researchers. These low cost sensors, like the Primesense sensor or the Kinect camera, acquire RGB data as normal camera and depth information simultaneously. These sensors are able to provide data for real-time applications. In this work, we present a plane detection algorithm based on PSO algorithm and the data obtained from an RGB-D sensor.

4.3.1 Description of the method

The algorithm begins with the initialization of the particles. Each particle is initialized with random values from -1 to 1 in order that they are inside the normalized plane. Then, for each particle x_i, we computed a candidate plane using the particle position and two neighbor particles. With these three points x_{i-1}, x_i, x_{i+1}, we computed the plane corresponding to each particle with

$$\pi = \left(\begin{array}{c} (x_{i-1} - x_{i+1}) \times (x_i - x_{i+1}) \\ -x_{i+1}^{\mathsf{T}}(x_{i-1} \times x_i) \end{array} \right).$$ (4.5)

With the candidate plane π, we computed the fitness value of the particle as

$$\eta_i = \sum_{i=0}^{W} \sum_{j=0}^{H} f(\pi \cdot x_{i,j}),$$ (4.6)

where W and H are the size of the window around the particle, and where

$$f(\alpha) = \begin{cases} 1 & \text{if } \alpha < \epsilon, \\ 0 & \text{otherwise,} \end{cases}$$ (4.7)

where ϵ is a threshold value. Then the local and global best values are computed, and the algorithm is repeated until a stop criterion is met.

The stop criterion can be defined as a maximum of iterations or a minimum of supporting plane points. The supporting points of a plane are the points that satisfy (4.7). A resume of the algorithm is shown in Algorithm 8.

4.3.2 Simulations results

The purpose of the simulations experiments was to test the behavior of the proposed approach under controllable varying factors. The simulation was conducted as follows: first, a 3D plane was defined, then its point-cloud computed. In the first test, we added different percentages of outliers. In

---

Algorithm 8 PSO algorithm for plane detection.

1: Randomly initialize particles with values from −1 to 1
2: **repeat**
3:     Compute a plane for each particle using (4.5)
4:     Compute fitness value η_i of each particle $\vec{x}_i$ using (4.6)
5:     Update local best b_i if the current fitness value η_i is better
6:     Determine global best γ
7:     **for all** particles **do**
8:         Update particle velocity $\vec{v}_i$
9:         Update particle position $\vec{x}_i$
10:     **end for**
11: **until** Stop criterion is met

---

the second set of experiments, we tested the proposed approach against different levels of noise. In the simulations, the reference plane was defined as $p = [0.19183, 0.408380, +0.892928, 2.13107]$.

4.3.2.1 Outliers test

In this experiment, we tested the proposed approach adding different percentages of outliers, varying from 10% to 90%. The results are presented in Table 4.1. For every outlier percentage level, the algorithm runs 1,000 times and outputs the estimated plane.

The results presented in Table 4.1 were obtained with a swarm size of 300 particles and a maximum of 30 iterations. We can observe that the estimated plane is the same as the reference plane. Therefore, we can conclude that the proposed approach is very robust to outliers.

With respect to the speed, it is notable that the proposed approach performs below 0.01, which can be enough for a real-time application. In regards to the 90% of outliers, it is obvious that the algorithm is slower compared to the other levels, but this is due to the amount of effort required to find the plane.

4.3.3 Noise test

The results presented in Table 4.2 were obtained with a swarm size of 300 particles and a maximum of 30 iterations. Different levels of noise were applied to the 3D point cloud. For every noise level, the algorithm ran 1,000 times and output the mean angle between the reference plane and the estimate plane, and the RMS distance between the planes.

Table 4.1 Outlier test. The percentage of outliers goes from 10% to 90%. The θ value represents the angle between the normal of the reference plane and the normal of the detected plane. The value D is the RMS of $d_r - d$, where d_r and d represent the distance of the plane from the origin of the reference plane and the estimated plane, respectively

% outlier	time (sec)	x	y	z	d	θ	D
10	0.000200	0.191830	0.408380	0.892428	2.131070	0.000000	0.000003
20	0.000400	0.191830	0.408380	0.892428	2.131070	0.000000	0.000003
30	0.000400	0.191830	0.408380	0.892428	2.131070	0.000000	0.000003
40	0.000600	0.191830	0.408380	0.892428	2.131070	0.000000	0.000003
50	0.000800	0.191830	0.408380	0.892428	2.131070	0.000000	0.000003
60	0.000800	0.191830	0.408380	0.892428	2.131070	0.000000	0.000003
70	0.001000	0.191830	0.408380	0.892428	2.131070	0.000000	0.000003
80	0.001200	0.191830	0.408380	0.892428	2.131070	0.000000	0.000003
90	0.463600	0.191830	0.408380	0.892428	2.131070	0.000000	0.000003

Table 4.2 Noise test. The percentage of noise goes from 5% to 50%. The θ value represents the mean angle between the normal of the reference plane and the normal of the detected plane. The value D represent the RMS of $d_r - d$, where d_r and d are the distance of the plane from the origin of the reference plane and the estimated plane, respectively

% noise	time (sec)	θ	D
5	0.0006	1.8513	0.0521
10	0.0002	3.2896	0.1102
15	0.0022	9.5670	0.2424
20	0.0036	10.6901	0.2479
25	0.0024	9.3214	0.3117
30	0.0028	9.0353	0.3792
35	0.0244	9.2277	0.6573
40	0.687	8.7435	0.8402
45	0.6624	8.8801	1.0338
50	0.7738	9.2361	1.2513

From these results, it is evident that large values of noise affected the depth more than the normal of the plane. For noise levels lower than 10%, the angle error is lower than 3.3 deg, and the distance between the planes is lower than 0.12 m. It is relevant to note that during the experiments with real data, the noise found was lower than 10%.

4.3.4 Experiments

The experiments were performed using a Primesense sensor, which is an RGB-D sensor. The sensor projects a pattern, which is captured by an infrared sensor to perform a depth estimation. The depth data is correlated with RGB-D camera. Therefore, we can obtain depth information for each image pixel.

The algorithm uses the RGB-D image to perform the search of the planes. Then a feasible plane is constructed using the depth information. The process is repeated until the plane with more supporting features is found. An example of the RGB-D data is presented in Fig. 4.4.

4.4. CONCLUSION

In this chapter, we presented a method to solve computer vision problems using soft computing algorithms. In particular, we showed how

Figure 4.4 RGB-D data figures.

to solve image tracking and plane detection using the particle swarm optimization.

The image tracking was performed using the PSO particles to search for the best template match. The objective function was defined using NCC. In order to deal with large images, we performed the search using an image pyramid. Therefore, the proposed algorithm works with the smallest image, improving the speed of the template search.

The plane detection algorithm uses the data provided by an RGB-D sensor. This data contains an RGB image and its corresponding depth information. The algorithm performs its search in the image plane, but it uses the depth information to construct the planes. The algorithm chooses the planes with more support features.

As evident from the presented applications, the soft computing algorithms are able to deal with computer vision problems. These examples suggest that the soft computing algorithms are a good option to solve computer vision problems.

REFERENCES

[1] Kennedy J, Eberhart R. Particle swarm optimization. In: Proceedings of the 1995 IEEE international conference on neural network; 1995.
[2] Lewis JP. Fast normalized cross-correlation. http://citeseerx.ist.psu.edu/viewdoc/summary?doi=10.1.1.21.6062, 1995.

CHAPTER FIVE

Soft Computing Applications in Mobile Robotics

Contents

5.1. INTRODUCTION TO MOBILE ROBOTICS

Manipulator robots are very important in the manufacturing industry [2]. These robots can perform many repetitive tasks with great speed and accuracy.

Regardless of their success, manipulator robots suffer from a fundamental disadvantage, the lack of mobility. In contrast, a mobile robot can move from one point to another to achieve its goal [3].

The main goal of robot navigation is to move a robot from its current position to a desired position, using the information provided by its sensors.

In general, robot navigation may be said to have two main tasks: one is global navigation and the other is local navigation. In the former case, the algorithms are related to path planning, i.e., having in the design the global path, from the start to the goal. In the latter case, the algorithms are related with the motion of the robot which must follow the global path as it detects and avoids obstacles.

In this chapter, we present a soft computing algorithm for local navigation that can be used with nonholonomic and holonomic robots.

Bio-inspired Algorithms for Engineering
https://doi.org/10.1016/B978-0-12-813788-8.00005-6

© 2018 Elsevier Inc.
All rights reserved.

5.2. NONHOLONOMIC MOBILE ROBOT NAVIGATION

In this work, the algorithm uses laser range finder readings. This sensor emits infra-red laser light and measures the distance to the object. The laser readings are defined in terms of polar coordinates and then converted into Cartesian coordinates. The readings that have a value equal or greater than the maximum range of the sensors are discarded since they do not represent a valid obstacle.

5.2.1 2D projective geometry

In this subsection, we present some properties of the projective geometry that are useful for facilitating the local navigation of a mobile robot.

We use 2D projective geometry to represent 2D points, lines, the intersection of lines, and the adjacency of points in a line.

5.2.1.1 Representation of points

In the Euclidean space, a 2D point can be represented as $\boldsymbol{x} = (a, b)^\top \in \mathcal{R}^2$. In addition, a 2D point can be represented by homogeneous coordinates, such as $\mathbf{x} = (x, y, w)^\top \in \mathcal{P}^2$, where vectors that differ by scale are considered to be equivalent, the space $\mathcal{P}^2$ is called the projective space.

A homogeneous vector can be converted back to a nonhomogeneous vector by dividing the vector with the last element, that is,

$$\boldsymbol{x} = \begin{pmatrix} x/w \\ y/w \\ 1 \end{pmatrix}. \tag{5.1}$$

When $w = 0$, the point is called point at infinity or ideal point. This point does not have an equivalent in the Euclidean space.

5.2.1.2 Representation of lines

In the projective space, a 2D line, $ax + bx + c = 0$ is defined as

$$\mathbf{1} = \begin{pmatrix} a \\ b \\ c \end{pmatrix}. \tag{5.2}$$

A point $\mathbf{x}$ in the line $\mathbf{1}$ satisfies

$$\mathbf{x} \cdot \mathbf{1} = 0, \tag{5.3}$$

a line $\mathbf{l}$ can be computed from two points $\mathbf{x}_1, \mathbf{x}_2$ with

$$\mathbf{l} = \mathbf{x}_1 \times \mathbf{x}_2. \tag{5.4}$$

The intersection of two lines can be computed with

$$\mathbf{x} = \mathbf{l}_1 \times \mathbf{l}_2. \tag{5.5}$$

5.2.1.3 Point line distance

A line $\mathbf{l} = (a, b, c)^\top$ in the projective space is defined with a normal vector

$$\hat{n} = \frac{n}{\|n\|}, \tag{5.6}$$

where $n = (a, b)^\top$, and a scalar

$$d = \frac{c}{|n|} \tag{5.7}$$

that represents the distance to the origin.

5.2.2 Robot navigation

In this local navigation work, we focused on obstacle avoidance involving following a set of subgoals from a path generated by a path-planner. The objective of the path-planner is the generation of a path from the initial position to the final position. However, this path cannot take into account moving or new obstacles. For that reason, the robot requires a local planner to take into account changes within the environment, Fig. 5.1.

Once the subgoals are defined, we require a controller to move the robot from its current position to the next subgoal. The controller used in this work is similar to the controller used in [1].

5.2.2.1 Nonholonomic robot kinematics

In this work, we focus on a differential drive mobile robot which is a robot with nonholonomic constraints. This type of robot has two drive wheels mounted on a common axis; each wheel can move forward or backward independently. This type of robot can move to any point on a plane. However, since it cannot move sideways it has to make some compensating maneuvers to reach its goal.

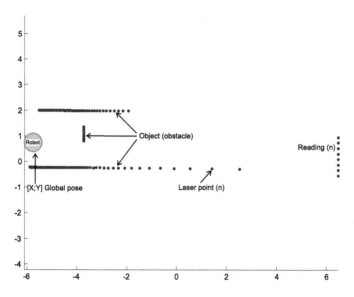

Figure 5.1 Robot and laser readings.

The kinematics of a differential drive mobile robot are defined as

$$\begin{pmatrix} \dot{x} \\ \dot{y} \\ \dot{\theta} \end{pmatrix} = \begin{pmatrix} \cos\theta & 0 \\ \sin\theta & 0 \\ 0 & 1 \end{pmatrix} \begin{pmatrix} v \\ \omega \end{pmatrix}, \tag{5.8}$$

where v, ω represent the linear, and angular velocities, respectively.

5.2.2.2 Feedback control

Let us assume, without loss of generality, that the goal is at the origin of the inertial frame, Fig. 5.2. Given the pose of the robot $\mathbf{p} = (x, y, \theta)^\top$, the error is defined as $\mathbf{e} = (x, y, \theta)^\top$.

The objective of the controller can be defined as

$$K\mathbf{e} = \begin{pmatrix} v \\ \omega \end{pmatrix}, \tag{5.9}$$

where

$$K = \begin{pmatrix} k_{1,1} & k_{1,2} & k_{1,3} \\ k_{2,1} & k_{2,2} & k_{2,3} \end{pmatrix} \tag{5.10}$$

drives the error $\mathbf{e}$ toward zero.

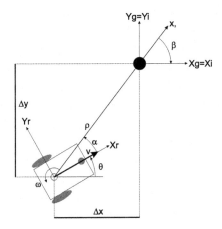

Figure 5.2 Robot kinematics.

The velocities v and ω must drive the robot from its current pose $(\rho_0, \alpha_0, \beta_0)$ to the goal position. The linear control can be defined as

$$v = \kappa_\rho \rho \qquad (5.11)$$

$$\omega = \kappa_\alpha \alpha + \kappa_\beta \beta, \qquad (5.12)$$

where

$$\rho = \sqrt{\Delta x^2 + \Delta y^2}, \qquad (5.13)$$

$$\alpha = \arctan 2(\Delta y, \Delta x) - \theta, \qquad (5.14)$$

$$\beta = -(\alpha + \theta). \qquad (5.15)$$

The system has a unique equilibrium point at $\mathbf{q} = (0, 0, 0)$. Therefore the controller will drive the robot to this point, the goal position [1].

5.2.3 Obstacle avoidance using PSO

The proposed approach is based on the PSO algorithm [4], where each particle represents a feasible position. The next position will be determined by the particle which is closest to the desired position and does not collide with an obstacle.

At each iteration, the PSO particles move around the robot minimizing the objective function. The objective function is defined as the Euclidean distance from the new position to the desired position. However, we consider only valid positions. That is, a position that does not collide with any

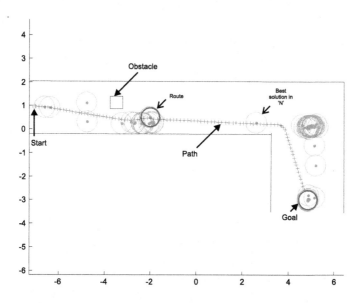

Figure 5.3 Robot navigation using PSO.

obstacle. If there is not any valid position, the PSO particles are randomly changed. The process is repeated until a valid position is found, Fig. 5.3.

5.2.3.1 Collision test

In this test we eliminate all the particles whose pose collides with an obstacle. To perform the test we define a circle with a radius bigger than the robot, the center of the circle is defined by each PSO particle.

To perform the collision test, we must check if the laser readings are inside or close to a circle. In those cases the poses are removed as indicated in Fig. 5.4.

5.2.3.2 Visibility test

In this test, we ensure that the feasible poses are visible, that is, the particles are not behind an obstacle. With this test, we check if the robot can travel from its current position to the desired position without a collision. All the particles that are not visible are discarded as shown in Fig. 5.5.

First, we define a line from the position of the robot $\mathbf{p} = (0, 0, 1)^\top$ and the desired position, $\mathbf{q}$. That is

$$\mathbf{l}' = \mathbf{p} \times \mathbf{q}. \tag{5.16}$$

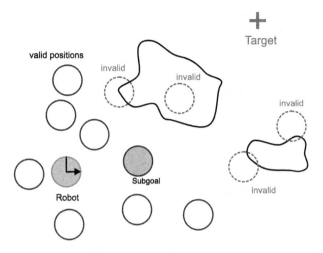

Figure 5.4 In the collision test, all the poses that collide with laser reading are discarded.

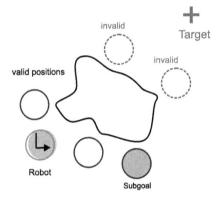

Figure 5.5 In the visibility test, all the poses that are behind and obstacle are discarded.

Then we normalize the line

$$\mathbf{l} = \frac{\mathbf{l}'}{\|\mathbf{l}'\|}. \tag{5.17}$$

With this line, $\mathbf{l} = (a, b, c)$, we define two new lines:

$$\mathbf{l}_1 = (a, b, \delta) \tag{5.18}$$

and

$$\mathbf{l}_2 = (a, b, -\delta). \tag{5.19}$$

We then define a line passing through the origin, and orthogonal to l, that is

$$l_3 = (b, -a, 0). \tag{5.20}$$

Then we define a line passing through the new position center $q = (x_d, y_d, 1)$ at a distance $d = \sqrt{x_d^2 + y_d^2}$. That is

$$l_4 = (-b, a, d). \tag{5.21}$$

To test if the proposed position is in a collision free path, we must check if there exist laser readings inside the rectangle defined with lines $l_1 \ldots l_4$. If such situation exists, it means that the desired position collides with an obstacle; therefore such position is discarded.

To test if the laser readings $r = (x, y, 1)^T$ is inside the rectangle, we used the dot product between the point and the lines. This can be defined as

$$inside(r) = step(p \cdot l_1) + step(p \cdot l_2) + step(p \cdot l_3) + step(p \cdot l_4). \tag{5.22}$$

If the value of $inside(r)$ is equal to 4, then the laser reading is inside the rectangle.

The function step is defined as

$$step(\alpha) = \begin{cases} 1 & \text{if } \alpha \geq 0 \\ 0 & \text{otherwise} \end{cases}. \tag{5.23}$$

5.2.3.3 Algorithm

A summary of the proposed approach is presented in Algorithm 9. The stop criterion is met when the proposed algorithm reached the total number of iterations.

5.3. HOLONOMIC MOBILE ROBOT NAVIGATION

Holonomic robots are robots that have the same controllable degrees of freedom as the degrees of freedom of the robot. In contrast, a nonholonomic robot has less controllable degrees of freedom than the degrees of freedom of the robot.

In this undertaking, we considered a robot with four Swedish wheels having rollers at 45°, each driven by a separate motor as shown in Fig. 5.6.

Algorithm 9 Algorithm of the proposed approach.

1:   Randomly initialize PSO particles

2:   **repeat**

3:       **repeat**

4:           Select particles that do not collide with obstacles

5:           Select particles that are not behind an obstacle

6:           Determine global best

7:           **if** No solution is found **then**

8:               Randomly initialize PSO particles around current robot pose

9:           **end if**

10:      **until** solution is found

11:      Move robot to subgoal

12:      **for all** particles **do**

13:          Update particle velocity $\mathbf{v}_i$

14:          Update particle position $\mathbf{x}_i$

15:      **end for**

16:   **until** Stop criterion is met

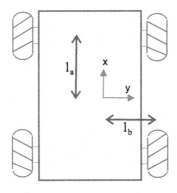

Figure 5.6 Omnidirectional mobile robot.

5.3.1 Kinematics of the holonomic robot

The omnidirectional mobile robot can spin around its vertical axis and move in any trajectory in the plane by varying the relative speed and the direction of rotation of the four wheels. Fig. 5.6 illustrates this kind of robot.

The kinematic model of the mobile robot is

$$
\begin{bmatrix} v_x \\ v_y \\ \omega \end{bmatrix} = R \begin{bmatrix} -\frac{1}{4} & \frac{1}{4} & -\frac{1}{4} & \frac{1}{4} \\ \frac{1}{4} & \frac{1}{4} & \frac{1}{4} & \frac{1}{4} \\ \frac{1}{4(l_a+l_b)} & -\frac{1}{4(l_a+l_b)} & -\frac{1}{4(l_a+l_b)} & -\frac{1}{4(l_a+l_b)} \end{bmatrix} \begin{bmatrix} \omega_1 \\ \omega_2 \\ \omega_3 \\ \omega_4 \end{bmatrix}, \quad (5.24)
$$

where R represents the radius of the wheel, and the values $\omega_1, \omega_2, \omega_3, \omega_4$ represent the angular velocity of each wheel. The value v_x denotes the velocity in the x axis (i.e. forward, backward movements), the value v_y denotes the velocity in the y axis (i.e. movement sideways). Finally, the value ω denotes the angular velocity of the robot.

5.3.1.1 Inverse kinematics

The inverse kinematics of the holonomic robot are defined as

$$
\begin{bmatrix} \omega_1 \\ \omega_2 \\ \omega_3 \\ \omega_4 \end{bmatrix} = \frac{1}{R} \begin{bmatrix} -1 & 1 & (l_a + l_b) \\ 1 & 1 & -(l_a + l_b) \\ -1 & 1 & -(l_a + l_b) \\ 1 & 1 & (l_a + l_b) \end{bmatrix} \begin{bmatrix} v_x \\ v_y \\ \omega \end{bmatrix}, \quad (5.25)
$$

where the inputs v_x, v_y and ω represent the driving velocity and the steering velocity, respectively. The value R denotes the radius of the wheel, the values l_a and l_b are defined in Fig. 5.6.

5.3.1.2 Obstacle avoidance

The algorithm used for obstacle avoidance is similar to Algorithm 9. The main difference is that the robot is able to move sideways. Therefore, in this case we use the velocity in the y direction, Fig. 5.7.

5.4. CONCLUSION

In this chapter, we presented a soft computing approach that is able to avoid obstacles while moving a robot to reach a goal. The approach is based on the PSO algorithm, where each particle represents a potential solution.

To select the best solution, we had to discard the particles that were not feasible solution. First, we defined a circle centered at the position of the particle. If there were laser readings inside the circle, or close to them, then the particle was discarded. Secondly, we checked if the particles were

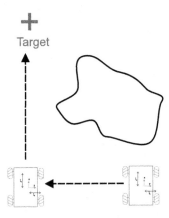

Figure 5.7 Omnidirectional robot obstacle avoidance.

visible to the robot, that is, they were not behind an obstacle. Any particle that was behind an obstacle was discarded.

Once we had discarded invalid particles, we chose the best particle. Then, we moved the robot to that position, and the process was repeated until we reached the goal. The algorithm has been tested with nonholonomic and holonomic robots.

As noted, the soft computing algorithms are able to solve local navigation problems in real-time and can be extended to solve other types of important robotics problems.

REFERENCES

[1] Astolfi A. Exponential stabilization of a mobile robot. In: European control conference; 1995. p. 3092–7. [Online]. Available from: http://control.ee.ethz.ch/index. cgi?page=publications;action=details;id=1188.

[2] Craig JJ. Introduction to robotics: mechanics and control. 4th ed. Pearson; 2017.

[3] Kelly A. Mobile robotics: mathematics, models, and methods. 1st ed. Cambridge University Press; 2013.

[4] Kennedy J, Eberhart R. Particle swarm optimization. In: Proceedings of IEEE international conference on neural networks; 1995.

CHAPTER SIX

Particle Swarm Optimization to Improve Neural Identifiers for Discrete-time Unknown Nonlinear Systems

Contents

6.1. INTRODUCTION

This chapter focuses on using particle swarm optimization to improve neural identifiers for discrete-time unknown nonlinear systems whose model is assumed to be unknown. These neural identifiers are robust in the presence of external and internal uncertainties. The first proposed scheme is based on a discrete-time Recurrent High Order Neural Network (RHONN) trained with a novel algorithm, based on Extended Kalman Filter (EKF) and Particle Swarm Optimization (PSO), using an on-line series-parallel configuration. The second proposed scheme is based on an EKF-improved algorithm using PSO to compute the design parameters. The EKF-PSO-based algorithm is employed to update the synaptic weights of the neural network. The proposed structure captures more efficiently the complex nature of wind speed and electricity prices time series, which are constantly monitored by a smart grid benchmark. For both schemes,

Bio-inspired Algorithms for Engineering
https://doi.org/10.1016/B978-0-12-813788-8.00006-8
© 2018 Elsevier Inc.
All rights reserved.
71

real-time results are included in order to illustrate the applicability of the proposed schemes.

 ## 6.2. PARTICLE-SWARM-BASED APPROACH OF A REAL-TIME DISCRETE NEURAL IDENTIFIER FOR LINEAR INDUCTION MOTORS

Linear Induction Motors (LIM) are special electrical machines in which the electrical energy is converted directly into mechanical energy of translatory motion. Strongest interest in these machines increased in the early 1970s. However, in the late 1970s, the research intensity and number of publications dropped. After 1980, LIM found their first notable applications in the transportation industry, automation, and home appliances, among others [9], [20].

LIM has many excellent performance features, such as high-starting thrust force, elimination of gears between motor and motion devices, reduction of mechanical losses, the size of motion devices, high speed operation, and silence [9], [46]. The driving principles of the LIM are similar to the traditional rotary induction motor (RIM), but their control characteristics are more complicated than the RIM. Their parameters are time varying, depending upon operating conditions, such as speed, temperature, and rail configuration.

Modern control systems usually require detailed knowledge about the system to be controlled. Such knowledge should be represented in terms of differential or difference equations. The mathematical description of the dynamic system is similar to that of the model.

There are various motives for establishing mathematical descriptions of dynamic systems. These include simulation, prediction, fault detection, and control system design. In this sense, basically there are two ways to obtain a model; it can be derived in a deductive manner using the laws of physics, or it can be inferred from a set of data collected during a practical experiment. The first method can be simple, but in many cases is excessively time-consuming. It would be unrealistic, or impossible, to obtain an accurate model in this way. The second method, typically referred to as system identification [51], could be a useful shortcut for deriving mathematical models. Although system identification does not always result in an accurate model, a satisfactory one can be often obtained with reasonable efforts. The main drawback is the requirement to conduct a practical experiment, which carry the system through its range of operation [17].

Due to their nonlinear modeling characteristics, neural networks have been successfully applied in control systems, pattern classification, pattern recognition, and identification problems. The best well-known training approach for recurrent neural networks (RNN) is the back propagation through time [21]. However, it is a first order gradient descent method and, hence, its learning speed could be very slow. Another well-known training algorithm is the Levenberg–Marquardt [21]. Its principal disadvantage is that there is no guarantee it will find the global minimum. Its learning speed could also be slow too, depending on the initialization.

In past years, EKF-based algorithms have been introduced to train neural networks [5]. With the EKF-based algorithm, learning convergence is improved [21]. The EKF training of neural networks, both feedforward and recurrent ones, have proven to be reliable for many applications [21]. However, EKF training requires the heuristic selection of some design parameters, which is not always an easy task [5], [2].

On the other hand, PSO technique, which is based on the behavior of a flock of birds or school of fish, is a type of evolutionary computing technique [27]. It has been shown that the PSO training algorithm takes fewer computations and is faster than the BP algorithm for neural networks to achieve the same performance [27].

In this chapter, a RHONN is used to design the proposed neural identifier for nonlinear systems whose mathematical model is assumed to be unknown. The learning algorithm for the RHONN is implemented using an EKF-PSO-based algorithm. Also herein, a class of Multi-Input Multi-Output (MIMO) discrete-time nonlinear system is considered for which a neural identifier is developed [38]; this identifier is then applied to a discrete-time unknown nonlinear system. The identifier is based on a RHONN [40] trained with an EKF-PSO-based algorithm. The applicability of these schemes is illustrated experimentally for a LIM [2].

6.2.1 Preliminaries

Throughout this chapter, k is used as the sampling step, $k \in \mathbb{N}$, $|\cdot|$ as the absolute value, and $\|\cdot\|$ as the Euclidean norm for vectors and as any adequate norm for matrices.

Consider an MIMO nonlinear system

$$\chi(k+1) = F(\chi(k), u(k)), \tag{6.1}$$
$$y(k) = h(\chi(k)), \tag{6.2}$$

where $\chi \in \Re^n$, $u \in \Re^m$, $y \in \Re^p$, and $F \in \Re^n \times \Re^m \to \Re^n$ and $h \in \Re^n \to \Re^p$ are nonlinear functions.

6.2.1.1 Discrete-time Recurrent High Order Neural Networks

The use of multilayer neural networks is well known for pattern recognition and for modeling of static systems. The NN is trained to learn an input–output map. Theoretical works have proven that, even with just one hidden layer, an NN can uniformly approximate any continuous function over a compact domain, provided that the NN has a sufficient number of synaptic connections [12], [41].

For control tasks, extensions of the first-order Hopfield model, RHONN, which present more interactions among the neurons, are proposed in [42]. Additionally, the RHONN model is very flexible and allows incorporation into the neural model, a priori information regarding the system's structure. Besides, discrete-time neural networks are better fitted for real-time implementations [5].

Consider the following discrete-time Recurrent High Order Neural Network (RHONN):

$$x_i(k+1) = w_i^\top z_i(x(k), u(k)), \quad i = 1, \cdots, n, \tag{6.3}$$

where x_i $(i = 1, 2, \cdots, n)$ is the state of the ith neuron, L_i is the respective number of higher-order connections, $\{I_1, I_2, \cdots, I_{L_i}\}$ is a collection of nonordered subsets of $\{1, 2, \cdots, n\}$, n is the state dimension, w_i $(i = 1, 2, \cdots, n)$ is the respective on-line adapted weight vector, and $z_i(x(k), u(k))$ is given by

$$z_i(x(k), u(k)) = \begin{bmatrix} z_{i_1} \\ z_{i_2} \\ \vdots \\ z_{i_{L_i}} \end{bmatrix} = \begin{bmatrix} \Pi_{j \in I_1} y_{i_j}^{d_{i_j}(1)} \\ \Pi_{j \in I_2} y_{i_j}^{d_{i_j}(2)} \\ \vdots \\ \Pi_{j \in I_{L_i}} y_{i_j}^{d_{i_j}(L_i)} \end{bmatrix} \tag{6.4}$$

with $d_{i_j}(k)$ being nonnegative integers, and y_i is defined as follows:

$$y_i = \begin{bmatrix} y_{i_1} \\ \vdots \\ y_{i_n} \\ y_{i_{n+1}} \\ \vdots \\ y_{i_{n+m}} \end{bmatrix} = \begin{bmatrix} S(x_1) \\ \vdots \\ S(x_n) \\ u_1 \\ \vdots \\ u_m \end{bmatrix}. \tag{6.5}$$

In (6.5), $u = [u_1, u_2, \ldots, u_m]^\top$ is the input vector to the neural network, and $S(\cdot)$ is defined by

$$S(x) = \frac{1}{1 + \exp(-\beta x)}. \tag{6.6}$$

Consider the problem to approximate the ith plant state for the general discrete-time nonlinear system (6.1) by the following discrete-time RHONN [19]:

$$\chi_i(k+1) = w_i^{*\top} z_i(x(k), u(k)) + \epsilon_{z_i}, \quad i = 1, \cdots, n, \tag{6.7}$$

where χ_i is the ith plant state, ϵ_{z_i} is a bounded approximation error, which can be reduced by increasing the number of the adjustable weights [42]. Assume that there exists ideal unknown weights vector w_i^* such that $\|\epsilon_{z_i}\|$ can be minimized on a compact set $\Omega_{z_i} \subset \Re^{L_i}$. The ideal weight vector w_i^* is an artificial quantity required only for analytical purposes and is defined as

$$w_i^* = \arg\min_{w_i} \left\{ \sup_{\chi, u} \left| F_i(\chi(k), u(k)) - w_i^\top z_i(\cdot) \right| \right\}.$$

It is assumed to be unknown, and it is the optimal set which renders the minimum approximation error, defined as ϵ_{z_i}; $F_i(\cdot)$ is the ith component of $F(\cdot)$ [17], [42]. Let us define its estimate as w_i and the estimation error as

$$\tilde{w}_i(k) = w_i^* - w_i(k). \tag{6.8}$$

Due to this fact, we use $w_i(k)$ as the approximation of the weight vector w_i^*, and ϵ_{z_i}, the modeling error, corresponds to $w_i^* \neq w_i(k)$. The estimate w_i is used for stability analysis, which will be discussed later. Since w_i^* is constant, $\tilde{w}_i(k+1) - \tilde{w}_i(k) = w_i(k+1) - w_i(k), \forall k \in \mathbb{N}$.

6.2.1.2 The EKF training algorithm

It is well known that KF estimates the state of a linear system with state and output additive white noises [10]. For KF-based neural network training, the network weights become the states to be estimated. In this case, the error between the neural network output and the measured plant output can be considered as the additive white noise [21]. Although the white noise assumption is seldom satisfied, the developed algorithm has proven to be efficient in real applications [5], [21]. Due to the fact that the neural network mapping is nonlinear, an EKF-type of training is required [40]. The training goal is to find the weight values that minimize the prediction error. In this chapter, an EKF-based training algorithm is used as follows:

$$
\begin{aligned}
w_i\left(k+1\right) &= w_i\left(k\right) + \eta_i K_i\left(k\right) e\left(k\right), & (6.9)\\
K_i\left(k\right) &= P_i\left(k\right) H_i\left(k\right) M_i\left(k\right),\\
P_i\left(k+1\right) &= P_i\left(k\right) - K_i\left(k\right) H_i^{\top}\left(k\right) P_i\left(k\right) + Q_i\left(k\right),\\
i &= 1,\cdots,n,
\end{aligned}
$$

with

$$
\begin{aligned}
M_i\left(k\right) &= \left[R_i\left(k\right) + H_i^{\top}\left(k\right) P_i\left(k\right) H_i\left(k\right)\right]^{-1}, & (6.10)\\
e\left(k\right) &= y\left(k\right) - \widehat{y}\left(k\right), & (6.11)
\end{aligned}
$$

where $e \in \Re^p$ is the output estimation error, $P_i \in \Re^{L_i \times L_i}$ is the weight estimation error covariance matrix at step k, $w_i \in \Re^{L_i}$ is the weight (state) vector, L_i is the respective number of neural network weights, $y \in \Re^p$ is the plant output, $\widehat{y} \in \Re^p$ is the NN output, n is the number of states, $K_i \in \Re^{L_i \times p}$ is the Kalman gain matrix, $Q_i \in \Re^{L_i \times L_i}$ is the NN weight estimation noise covariance matrix, $R_i \in \Re^{p \times p}$ is the error noise covariance, and $H_i \in \Re^{L_i \times p}$ is a matrix in which each entry (H_{ij}) is the derivative of the ith neural output with respect to ijth neural network weight (w_{ij}), given as follows:

$$
H_{ij}\left(k\right) = \left[\frac{\partial y\left(k\right)}{\partial w_{ij}\left(k\right)}\right]^{\top}, \tag{6.12}
$$

where $i = 1, ..., n$ and $j = 1, ..., L_i$. Usually P_i, Q_i, and R_i are initialized as diagonal matrices with entries $P_i(0)$, $Q_i(0)$, and $R_i(0)$, respectively. Given that typically the entries $P_i(0)$, $Q_i(0)$, and $R_i(0)$ are defined heuristically, in this chapter, we propose the use of a PSO algorithm in order to compute on-line such entries to improve the EKF training algorithm, as explained below.

6.2.1.3 PSO improvement for EKF training algorithm

To improve the performance of the basic PSO algorithm, some new versions of it have been proposed. At first, the concept of an inertia weight was developed to better control exploration and exploitation in [27], [44], [52]. Then, the research done by Clerc [13] indicated that using a constriction factor may be necessary to ensure convergence of the particle swarm algorithm. After these two important modifications in the basic PSO were introduced, the Multi-Phase Particle Swarm Optimization (MPSO), the PSO-Gaussian mutation, the quantum particle swarm optimization, a modified PSO with increasing inertia weight schedule, the Gaussian particle swarm optimization (GPSO), and the guaranteed convergence PSO (GCPSO) were all introduced in [1]. In this section, the PSO algorithm explained in Section 1.2 of Chapter 1 is used in order to determine the design parameters for the EKF-Learning algorithm.

Initially, a set of random solutions or a set of particles are considered. A random velocity is given to each particle, and they are flown through the problem space. Each particle has memory, which is used to keep track of the previous best position and corresponding fitness. The best value of the position of each individual is stored as p_{id}. In other words, p_{id} is the best position acquired by a particle during the course of its movement within the swarm. It has another value called the $\%p_{gd}$, which is the best value of all the particles p_{id} in the swarm. The basic concept of the PSO technique lies in accelerating each particle towards its p_{id} and p_{gd} locations at each time step. The PSO algorithm used in this chapter is depicted in Fig. 6.1 and can be defined as follows [27]:

1. Initialize a population of particles with random positions and velocities in the problem space;
2. For each particle, evaluate the desired optimization fitness function;
3. Compare the particles fitness evaluation with the particles p_{id} if the current value is better than the p_{id}, then set p_{id} value equal to the current location;
4. Compare the best fitness evaluation with the population's overall previous best. If the current value is better than the p_{gd}, then set $\%p_{gd}$ to the particle's array and index value;
5. Update the particle's velocity and position as follows:
   The velocity of the ith particle of d dimension is given by

$$v_{id}\left(k+1\right) = c_0 v_{id}\left(k\right) + c_1 \operatorname{rand}_1\left(p_{id}\left(k\right) - x_{id}\left(k\right)\right)$$
$$+ c_2 \operatorname{rand}_2\left(p_{gd}\left(k\right) - x_{id}\left(k\right)\right).$$

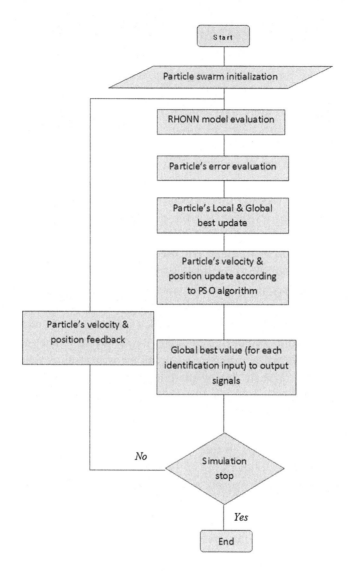

Figure 6.1 PSO algorithm.

The position vector of the ith particle of d dimension is updated as follows:

$$x_{id}\left(k+1\right) = x_{id}\left(k\right) + v_{id}\left(k\right),$$

where c_0 is the inertia weight, c_1 is the cognition acceleration constant, and c_2 is the social acceleration constant.

6. Repeat step 2 until a criterion is met, usually a sufficiently good fitness or a maximum number of iterations or epochs.

If the velocity of the particle exceeds V_{max} (the maximum velocity for the particles), then it is reduced to V_{max}. Thus, the resolution and fitness of search depends on the V_{max}. If V_{max} is too high, then particles will move in larger steps, and so the solution reached may not be as good as expected. If V_{max} is too low, then particles will take a long time to reach the desired solution [27]. Due the above explained, PSO are very suitable models for noisy problems, just as the one we are considering. Given that PSO has shown good results in optimization problems [27], it will be used to optimize the values for KF covariance matrices instead of heuristic solutions. For this purpose, each particle will represent one of the Kalman's covariance entries.

6.2.2 Neural identification

Literature reports the capacity of RHONN to identify nonlinear systems in continuous-time [42] as well as in discrete-time [5], [6]. However, the accuracy of results is greatly dependent on the RHONN structure selection, as well as its training. For real-time implementations, RHONN—in discrete time—has shown excellent results. However, as it is demonstrated in [28], the identification error can be minimized by increasing the number of high-order connections, although it is not possible to eliminate it in real-life problems. On the other hand, it is possible to reduce the identification error by means of the selection of an adequate training algorithm. Nonetheless, the EKF training algorithm has proven to be reliable for many applications, particularly for real-time implementations [6], [21]. However, EKF training requires the heuristic selection of some design parameters, which is not always an easy task. Besides, the adequate selection of the design parameter directly affects the bound of the identification error [6]. Therefore a systematic methodology to select the design parameters is an important contribution for neural identification of unknown discrete-time nonlinear systems.

Now, the RHONN (6.7) is trained with an EKF-PSO algorithm, as defined above, to identify the nonlinear system (6.3). First, a RHONN structure is proposed. Such structure can have physical significance or not. For control tasks it is better to consider a RHONN structure with physical significance. Then, the EKF (6.9)–(6.11) is selected to implement the on-line series parallel training of the RHONN. However as mentioned above,

EKF training algorithm requires adequate selection of many design parameters, particularly covariance matrices and learning rate. The latter directly affect the identification error, and it is hard to determine their suitable values. In order to simplify the training algorithm tuning and to improve the identification process, the use of standard PSO algorithm (Fig. 6.1) to determine such parameters is proposed in this chapter. Finally, an improved identification scheme for unknown discrete-time nonlinear systems is obtained.

6.2.3 Linear Induction Motor application

In this section, the above developed scheme is applied to identify a three-phase linear induction motor. It is important to note that the proposed scheme is developed assuming that the plant model, that is, parameters as well as external disturbances (load torque) are unknown [2].

6.2.3.1 Motor model

In order to illustrate the applicability of the proposed scheme, in this section, the proposed neural identifier is applied to the $\alpha - \beta$ model of an LIM discretized by the Euler technique, which—for the purposes of this chapter—is considered unknown [8], [25], [33] as follows:

$$q_m(k+1) = q_m(k) + v(k)T,$$

$$v(k+1) = (1 - K_2 T)v(k) - k_1 T\lambda_{r\alpha}(k)\rho_1 i_{s\alpha}(k) - k_1 T\lambda_{r\beta}(k)\rho_2 i_{s\alpha}(k)$$
$$+ k_1 T\lambda_{r\alpha}(k)\rho_2 i_{s\beta}(k) - k_1 T\lambda_{r\beta}(k)\rho_1 i_{s\beta}(k) - k_3 TF_L,$$

$$\lambda_{r\alpha}(k+1) = (1 - k_6 T)\lambda_{r\alpha}(k) + k_4 Tv(k)\rho_1 i_{s\alpha}(k) - k_4 T\rho_1 i_{s\alpha}(k) + k_5 T\rho_2 i_{s\alpha}(k)$$
$$+ k_4 T\rho_2 i_{s\beta}(k) - k_4 Tv(k)\rho_2 i_{s\beta}(k) + k_5 T\rho_1 i_{s\beta}(k),$$

$$\lambda_{r\beta}(k+1) = (1 - k_6 T)\lambda_{r\beta}(k) + k_4 Tv(k)\rho_2 i_{s\alpha}(k) - k_4 T\rho_2 i_{s\alpha}(k) - k_5 T\rho_1 i_{s\alpha}(k)$$
$$+ k_4 T\rho_1 i_{s\beta}(k) + k_4 Tv(k)\rho_1 i_{s\beta}(k) + k_5 T\rho_2 i_{s\beta}(k),$$

$$i_{s\alpha}(k+1) = (1 + k_9 T)i_{s\alpha}(k) - k_7 T\lambda_{r\alpha}(k)\rho_2 - k_8 T\lambda_{r\alpha}(k)v(k)\rho_1$$
$$+ k_7 T\lambda_{r\beta}(k)\rho_1 - k_8 T\lambda_{r\beta}(k)v(k)\rho_2 - k_{10} Tu_\alpha(k),$$

$$i_{s\beta}(k+1) = (1 + k_9 T)i_{s\beta}(k) + k_8 T\lambda_{r\alpha}(k)v(k)\rho_2 - k_7 T\lambda_{r\alpha}(k)\rho_1$$
$$- k_7 T\lambda_{r\beta}(k)\rho_2 - k_8 T\lambda_{r\beta}(k)v(k)\rho_1 - k_{10} Tu_\beta(k), \qquad (6.13)$$

with

$$\rho_1 = \sin(n_p q_m(k)), \qquad \rho_2 = \cos(n_p q_m(k)),$$

$$k_1 = \frac{n_p L_{sr}}{D_m L_r}, \qquad k_2 = \frac{R_m}{D_m},$$

$$k_3 = \frac{1}{D_m}, \qquad k_4 = n_p L_{sr},$$

$$k_5 = \frac{R_r L_{sr}}{L_r}, \qquad k_6 = \frac{R_r}{L_r},$$

$$k_7 = \frac{L_{sr} R_r}{L_r(L_{sr}^2 - L_s L_r)}, \qquad k_8 = \frac{L_{sr} n_p}{L_{sr}^2 - L_s L_r},$$

$$k_9 = \frac{L_r^2 R_s + L_{sr}^2 R_r}{L_r(L_{sr}^2 - L_s L_r)}, \qquad k_{10} = \frac{L_r}{L_{sr}^2 - L_s L_r},$$

where q_m is the position, v is the linear velocity, $\lambda_{r\alpha}$ and $\lambda_{r\beta}$ are the α and β secondary flux components, respectively, $i_{s\alpha}$ and $i_{s\beta}$ are the α and β primary current components, respectively, $u_{s\alpha}$ and $u_{s\beta}$ are the α and β primary voltage components, respectively, R_s is the winding resistance per phase, R_r is the secondary resistance per phase, L_{sr} is the magnetizing inductance per phase, L_s is the primary inductance per phase, L_r is the secondary inductance per phase, F_L is the load disturbance, D_m is the viscous friction and iron-loss coefficient, n_p is the number of poles pairs, and T is the sampling period [8].

It is important to note that this mathematical model is considered unknown for the design of the neural identifier. It is only included in this chapter for completeness purposes [2].

6.2.3.2 Neural identifier design

The neural identifier proposed is designed as follows:

$$x_1(k+1) = w_{11}(k) S(v(k)) + w_{12}(k) S(q_m(k)),$$

$$x_2(k+1) = w_{21}(k) S(v(k))^2 + w_{22}(k) S(\lambda_{r\alpha}(k))^2 + w_{23}(k) S(\lambda_{r\beta}(k))^{15},$$

$$x_3(k+1) = w_{31}(k) S(v(k))^2 + w_{32}(k) S(\lambda_{r\alpha}(k))^2 + w_{33}(k) S(\lambda_{r\beta})^{15},$$

$$x_4(k+1) = w_{41}(k) S(v(k))^2 + w_{42}(k) S(\lambda_{r\alpha}(k))^2 + w_{43}(k) S(\lambda_{r\beta}(k))^{15},$$

$$x_5(k+1) = w_{51}(k) S(v(k))^2 + w_{52}(k) S(\lambda_{r\alpha}(k))^2 + w_{53}(k) S(\lambda_{r\beta}(k))^2$$
$$+ w_{54}(k) S(i_{s\alpha}(k))^3 + 0.02178 u_\alpha(k),$$

$$x_6(k+1) = w_{61}(k) S(v(k))^2 + w_{62}(k) S(\lambda_{r\alpha}(k))^2 + w_{63}(k) S(\lambda_{r\beta}(k))^2$$
$$+ w_{64}(k) S(i_{s\beta}(k))^3 + 0.02178 u_\beta(k), \qquad (6.14)$$

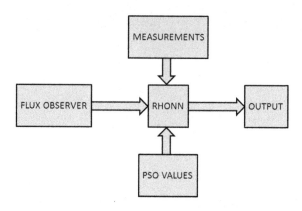

Figure 6.2 LIM identification scheme.

where $S(x(k)) = \alpha \tan h(\beta x\,(k))$, $x_1(k)$ to identify $q_m(k)$, $x_2(k)$ to identify $v(k)$, $x_3(k)$ to identify $\lambda_{r\alpha}(k)$, $x_4(k)$ to identify $\lambda_{r\beta}(k)$, $x_5(k)$ to identify $i_{s\alpha}(k)$, and $x_6(k)$ to identify $i_{s\beta}(k)$. For this application, only the fluxes are considered immeasurable. The training is performed on-line using a series-parallel configuration as shown in Fig. 6.2. Both the NN and LIM states are initialized randomly. The associated covariance matrices are computed using the PSO algorithm, and the RHONN weights are updated with the EKF as in (6.9). The input signals u_α and u_β are selected as chirp functions.

Reduced order nonlinear observer

The proposed neural identifier (6.14) requires the full state measurement assumption [8]. However, for real-time implementations, rotor fluxes measurement is a difficult task. Here, a reduced order nonlinear observer is designed for fluxes on the basis of rotor speed and current measurements. The flux dynamics is demonstrated in (6.13). Therefore the following observer is used [22]:

$$\tilde{\Lambda}(k+1) = \tilde{\Lambda}(k) - k_6\,T\tilde{\Lambda}(k) - k_4\,T\Theta^T JI_s(k)$$
$$+ k_4\,T\Theta^T JI_s(k)v(k) + k_4\,T\Theta^T I_s(k), \qquad (6.15)$$

where

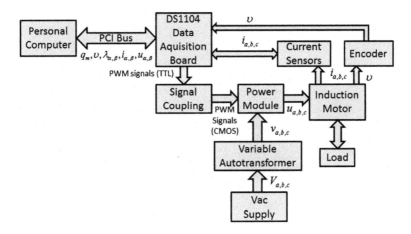

Figure 6.3 Schematic representation of the prototype to be identified.

$$
\Lambda(k) = \begin{bmatrix} \lambda_{r\alpha}(k) \\ \lambda_{r\beta}(k) \end{bmatrix},
$$

$$
I_s(k) = \begin{bmatrix} i_{s\alpha}(k) \\ i_{s\beta}(k) \end{bmatrix},
$$

$$
\Theta(k) = \begin{bmatrix} \cos(n_p q_m(k)) & -\sin(n_p q_m(k)) \\ \sin(n_p q_m(k)) & \cos(n_p q_m(k)) \end{bmatrix},
$$

$$
J = \begin{bmatrix} 0 & -1 \\ 1 & 0 \end{bmatrix}.
$$

The stability proof for (6.15) is presented in [22].

6.2.3.3 Experimental results

The proposed scheme is depicted in Fig. 6.2. The experiments are performed using a benchmark whose schematic representation is depicted in Fig. 6.3. Fig. 6.4 shows the experimental benchmark for the LIM.

The methodology used to implement the experimental identifier is as follows:

1. Validate and test this algorithm via simulation in Matlab/Simulink using a plant model and their respective parameters;
2. Download the validated identifier to the DS1104 board;

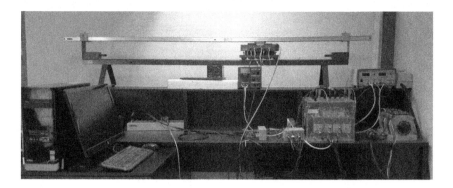

Figure 6.4 Linear induction motor prototype.

3. Replace the simulated model state variable values by the induction motor measurements (current and angular position), acquired through the DS1104 board A/D ports, and calculated (fluxes) state variables values;
4. Send back—through the DS1104 board—the input signals (voltages) defined as chirp signals;
5. Process the input signals through the Space Vector Pulse Width Modulation (SVPWM) power stage; and
6. Apply the SVPWM output to the induction motor.

Experimental results are presented as follows. Fig. 6.5 displays the identification performance detail for position. It is obvious that noise exists in the plant signal. However, RHONN is capable to identify position signal. Fig. 6.6 illustrates the identification performance for linear velocity. It is possible to note that identification errors remain bounded and decays faster to a minimum value. Fig. 6.7 and Fig. 6.8 present the identification performance for the fluxes in phase α and β, respectively. For both variables the identification is reached at the minimum of 100 ms with an adequate accuracy. Figs. 6.9 and 6.10 portray the identification performance for currents in phases α and β, respectively. From both figures, it is possible to appreciate that a suitable identification is performed. However, evolution of each one is different due to many circumstances, as well as the structure selected for each neural state variable (6.14), and/or the presence of external, and internal disturbances, not necessarily equal for each component. Finally, Fig. 6.11 shows the identification errors, it can be seen that all of them remain bounded. Besides, from Fig. 6.11, it is possible to note that the bounds for identification errors for each state variable are different. In

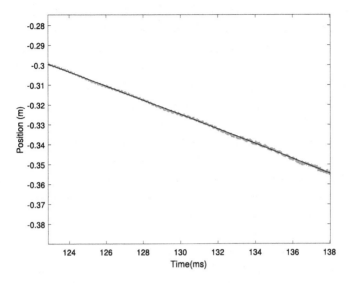

Figure 6.5 Position identification. Plant signal is in solid line and identified signal in dashed line.

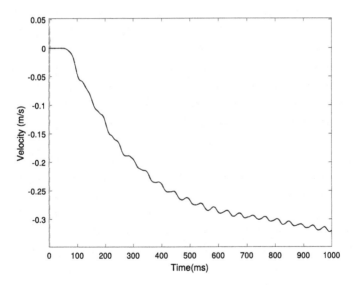

Figure 6.6 Velocity identification. Plant signal is in solid line and identified signal in dashed line.

fact they can be affected by internal and external disturbances of the system. In this particular experiment, it can be observed that the beta components for current and flux are more affected than the alpha ones. This can be an

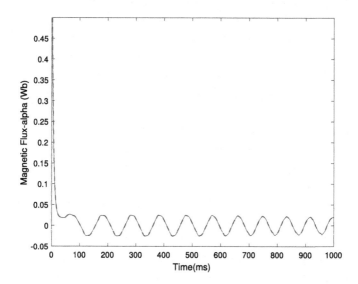

Figure 6.7 Alpha flux identification. Plant signal is in solid line and identified signal in dashed line.

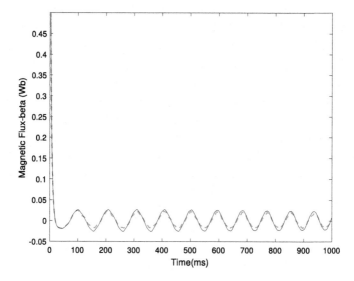

Figure 6.8 Beta flux identification: plant signal is in solid line and identified signal in dashed line.

instrumentation problem. However, in the case of neural network training, this problem can be seen as an opportunity. Due to the existence of internal and external disturbances the neural network can perform an ad-

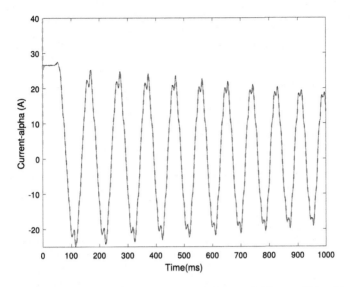

Figure 6.9 Alpha current identification: plant signal is in solid line and identified signal in dashed line

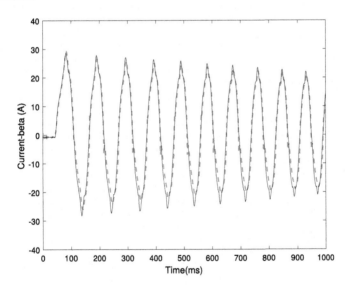

Figure 6.10 Beta current identification: plant signal is in solid line and identified signal in dashed line.

equate identification of the state variables and the inclusion of PSO can improve such identification scheme as explained below. It is important to consider that experimental results, depicted in Figs. 6.5–6.11, have been

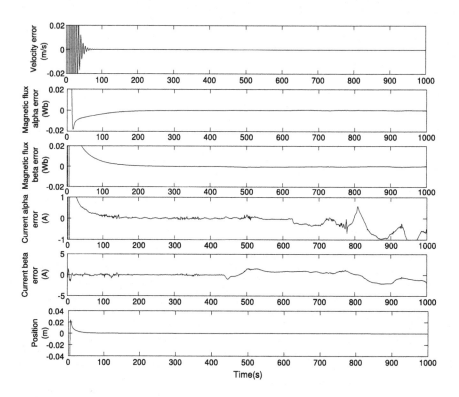

Figure 6.11 Identification errors.

obtained in open loop with chirp functions as inputs for LIM. In order to excite most of the plant dynamics, moreover, the neural states are initialized randomly.

6.2.3.4 Comparison of the EKF-PSO algorithm for neural identification

In order to evaluate the performance of the EKF-PSO algorithm, it is compared with the typical EKF algorithm [6]; both training algorithms are used to identify a LIM in real-time. The RHONN structure are exactly the same (6.14); only the training algorithm is changed in order to compare its performance.

Table 6.1 includes a comparison between the proposed PSO-EKF learning algorithm and the EKF one.

Table 6.1 results show that the proposed methodology leads to an improvement of the results compared with the EKF training algorithm.

Table 6.1 Mean value (μ) for identification error

Variable	EKF	PSO-EKF
ν	4.2271×10^{-4}	2.6469×10^{-4}
$\lambda_{r\alpha}$	6.2001×10^{-3}	2.8135×10^{-3}
$\lambda_{r\beta}$	1.5142×10^{-3}	4.2102×10^{-3}
$i_{s\alpha}$	4.0230×10^{-1}	5.7903×10^{-3}
$i_{s\beta}$	2.7529×10^{-1}	5.5211×10^{-3}
q	9.0127×10^{-3}	1.7351×10^{-4}

6.3. NEURAL MODEL WITH PARTICLE SWARM OPTIMIZATION KALMAN LEARNING FOR FORECASTING IN SMART GRIDS

The limited existing reserves of fossil fuel, and the harmful emissions associated with them, have led to an increased focus on renewable energy applications in recent years. The first steps toward integrating renewable energy sources began with hybrid wind and solar systems as complementing sources, and solution, for rural applications and weak grid interconnections. Further research have implemented hybrid systems, including several small scale renewable energy sources, such as solar, thermal, biomass, fuel cells, and tidal power. Since the production costs for photovoltaic and wind turbine applications have considerably reduced, they have become the primary choices for hybrid energy generation systems. The future of energy production is headed towards this scheme of integration of renewable energy sources with existing conventional generation systems characterized by a high degree of measurement, communications and control. This integration is defined as a smart grid. This new scheme increases the power quality since the production becomes decentralized. The latter is the main reason why institutions have increased research on this concept [34]. Microgrids integrate small scale energy generation systems mainly from renewable energy and the implementation of complex control technologies to improve the flexibility and reliability of power systems. That is, the design of these systems integrates a distributed power generation system and a management unit composed of a communication network which monitors and controls the interconnection between energy sources, storage devices, and electrical loads.

Among renewable energy sources, wind energy is the one with the lowest cost of electricity production [47]. However, in practice, the integration of wind energy into the existing electricity supply system is a real challenge because its availability mainly depends on meteorological conditions, particularly on the magnitude of the wind speed, which cannot directly be changed by human intervention. For this reason, it is important to have a reliable estimation of wind velocity and direction which directly impact energy generation. Integration of the wind speed forecast and output power is a good way to improve smart grids scheduling performance [23]. Wind prediction is not an easy task; the wind has a stochastic nature with high rate of change. Wind speed time series present highly nonlinear behavior with no typical patterns and a weak seasonal character [3].

Several methods have been proposed to accomplish wind characteristics forecasting, such as numerical weather prediction systems, statistical approaches, and artificial neural networks using feedforward or recurrent structures ([44], [16], [24], [43], [47], [50]). In [16], a linear time-series-based model relating the predicted interval to its corresponding one and data covering a temporal span of two years is developed. The statistical approaches have the advantage of low cost since they only require historical data. On the other hand, the accuracy of the prediction drops for long time horizons. Artificial intelligence methods are more suitable for short term predictions; these methods are based on time series historical data in order to build a mathematical model which approximates the input–output relationship. Time-series-based models include the autoregressive (AR) and the autoregressive moving average (ARMA) models, which predict the average wind speed for one step ahead [50].

Artificial Neural Networks (ANN) have been considered as a convenient analysis tool for the forecasting and control aspects of wind generation systems due to the simplicity of the model and the accuracy of the results for nonlinear and stochastic models. The ANN model has also been implemented in several practical applications [43].

In [30], a merged Neuro Fuzzy system is developed as a universal approximator in order to estimate the state of charge in a battery bank.

In [43], a Recurrent Neural Network is applied to forecast the output power of wind generators based on wind speed prediction using one year of historical data generated from hour-ahead to day-ahead predictions with errors ranging from 5% for one hour horizon to 20% for one day ahead forecasting. In [44], local recurrent neural networks are implemented to forecast wind speed and electrical power in a wind park with a seventy two hour ahead forecast. A better performance was obtained in comparison with static network approaches.

ANN have been previously implemented for wind power short term predictions. They outperformed other classical methods due to the fast learning algorithm which enables on-line implementations and the versatility to vary the prediction horizon [24]. Due to their nonlinear modeling characteristics, neural networks have been successfully applied in control systems, pattern classification, pattern recognition, and time series forecasting problems.

There are several previous works that use artificial neural networks to predict wind time series [3], [39], [47]. The best well-known training approach for recurrent neural networks (RNN) is the back propagation through time [49]. However, it is a first-order gradient descent method, and hence its learning speed could be very slow [29]. Another well-known training algorithm is the Levenberg–Marquardt algorithm [36]; its principal disadvantage is that its finding the global minimum is not guaranteed and its learning speed could be slow too, depending on the initialization. In past years, EKF-based algorithms were introduced to the training of neural networks [5], [18]. With the EKF based algorithm, the learning convergence is improved [29]. The EKF training of neural networks, both the feedforward and recurrent ones, have proven to be reliable for many applications over the past ten years [18]. However, EKF training requires the heuristic selection of some design parameters, which is not always an easy task [5], [39], [4].

During the past decade, the use of evolutionary computation in engineering applications has increased. Evolutionary algorithms apply adaptation stochastically in optimization problems, such as evolutionary programming, genetic algorithms and evolution strategies [37]. The PSO technique, which is based on the behavior of a flock of birds or school of fish, is a type of evolutionary computing technique [27]. The PSO algorithm uses a population of search points that evolve in a search space using a communication method to transfer the acquired experience from the best solutions. This algorithm has several advantages, including the simplicity of the updating law, faster convergence time, and less complexity on the reorganization of the population. The PSO methods have also emerged as excellent tools to improve the performance of neural network learning process [45]. In [32], a PSO learning rule for a neural network is implemented using FPGA for dynamic system identification. In [31], the PSO algorithm is extended to multiple swarms in a neuro-fuzzy network with good results in forecasting applications. It has been shown that the PSO training algorithm takes fewer computations and is faster than the BP algorithm for neural networks to achieve the same performance [27].

In this chapter, we propose the use of PSO for tuning the parameters of EKF training algorithm. This scheme is a new one and is suitable for data modeling in smart grids since the forecasting horizon satisfy the requirements for several applications in the grid operation. The length of the regression vector is determined using the Cao methodology which is an improvement to the false neighbors approach [11]. The applicability of this architecture is illustrated via simulation using real data values for wind speed. That is, in order to show the potential applications in forecasting for energy generation in smart grid schemes [4].

6.3.1 Neural identification

In this chapter, for the neural model identification, the Recurrent Multi-Layer Perceptron is chosen. As a result, the neural model structure problem reduces to dealing with the following issues: selecting the inputs to the network and selecting the internal architecture of the network.

The structure selected in this chapter is NNARX (acronym for Neural Network AutoRegressive eXternal input) [36]. The output vector for the artificial neural network is defined as the regression vector of an AutoRegressive eXternal input linear model structure (ARX) [5].

It is common to consider a general nonlinear system. However, for many control applications it is preferred to express the model in an affine form. The latter can be represented by the following equations:

$$y(k+1) = f(y(k), y(k-1), ..., y(k-q+1)),   (6.16)$$

where q is the dimension of the regression vector. In other words, there exists a nonlinear mapping f for which the present value of the output $y(k+1)$ is uniquely defined in terms of its past values $y(k), ..., y(k-q+1)$ and the present values of the input $u(k)$.

Considering that it is possible to define

$$\phi(k) = [y(k), ..., y(k-q+1)]^{T},$$

which is similar to the regression vector of an ARX linear model structure [36], the nonlinear mapping f can be approximated by a neural network defined as

$$y(k+1) = \varphi(\phi(k), w^{*}) + \varepsilon,$$

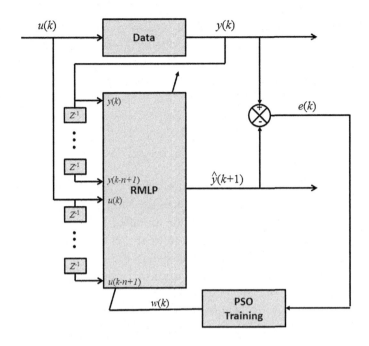

Figure 6.12 Neural network structure.

where w^* is an ideal weight vector, and ε is the modeling error. Such a neural network can be implemented on predictor form as

$$\widehat{y}(k+1) = \varphi\left(\phi\left(k\right), w\right), \qquad (6.17)$$

where w is the vector containing the adjustable parameters in the neural network.

The neural network structure, used in this work is depicted in Fig. 6.12, which contains sigmoid units only in the hidden layer; the output layer is a linear one. The used sigmoid function $S\left(\cdot\right)$ is defined as a logistic one as follows:

$$S(\varsigma) = \frac{1}{1 + \exp\left(-\beta\varsigma\right)}, \qquad \beta > 0, \qquad (6.18)$$

where ς is any real-valued variable.

6.3.1.1 EKF training algorithm improved with PSO

KF estimates the state of a linear system with additive state and output white noise. Kalman filter algorithm is developed for a linear, discrete-time dynamical system. For KF-based neural network training, the network weights become the states to be estimated. Due to the fact that the neural network mapping is nonlinear, an EKF-type is required [5].

Consider a nonlinear dynamic system described by the next model in state space

$$
\begin{aligned}
w(k+1) &= f(k, w(k)) + v_1(k), \\
y(k) &= h(k, w(k)) + v_2(k), \quad (6.19)
\end{aligned}
$$

where $v_1(k)$ and $v_2(k)$ are zero-mean white noises with covariance matrices represented by $Q(k)$ and $R(k)$, respectively, and $f(k, w(k))$ denotes the nonlinear transition matrix function. The modified Extended Kalman Filter (EKF) algorithm has been defined in (6.9)–(6.12) and the algorithm depicted in Fig. 6.1 is used in order to determine the design parameters for the EKF-Learning algorithm. Initially, a set of random solutions or a set of particles are considered. A random velocity is given to each particle and they are flown through the problem space. Each particle has memory, which is used to keep track of the previous best position and corresponding fitness.

6.3.1.2 Regressor structure

We now discuss the choice of an appropriate number of delayed signals to be used in the training phase.

A wrong number of delayed signals, used as regressors, could have a substantially negative impact on the training process, while a too small number implies that essential dynamics will not be modeled. Additionally, a large number of regression terms increases the required computation time. Also, if too many delayed signals are included in the regression vector, it will contain redundant information. For good behavior of the model structure, it is necessary to have both a sufficiently large lag space and an adequate number of hidden units. If the lag space is properly determined, the model structure selection problem is substantially reduced. There have been many discussions of how to determine the optimal embedding dimension from a scalar time series based on Takens' theorem [11]. The basic methods, which are usually used to choose the minimum embedding dimension, are: (1) computing some invariant on the attractor, (2) singular value decomposition, and (3) the method of false neighbors. However, a practical method to select the lag space is the one proposed by Cao [11] to determine the minimum embedding dimension. It overcomes most of the shortcomings of the above mentioned methodologies, like high dependence on design parameters and high computational cost, among others [11]. Besides, it has several advantages: it does not contain any subjective parameters except for the time-delay embedding; does not strongly depend on how many data points are available; can clearly distinguish deterministic signals from stochastic signals; works well for time series from high-dimensional attractors; and is computational efficient. In this chapter, this technique for determination of the optimal regressor structure is used.

Let us consider a time series $x_1, x_2, ..., x_n$ and define a set of time-delay vectors as

$$y_i = \begin{bmatrix} x_i & x_{i+\tau} & ... & x_{i+(d-1)\tau} \end{bmatrix},$$
$$i = 1, 2, ..., N - (d-1)\tau,$$

where d is the embedding dimension. This dimension is determined from the evolution of a function $E(d)$, defined as

$$E(d) = \frac{1}{N - d\tau} \sum_{i=1}^{N-d\tau} \frac{\left\| y_i(d+1) - y_{n(i,d)}(d+1) \right\|}{\left\| y_i(d) - y_{n(i,d)}(d) \right\|}$$
$$i = 1, 2, ..., N - d\tau$$

where $n(i, d)$ is an integer such that $y_{n(i,d)}(d)$ is the nearest neighbor of $y_i(d)$ [26]. The minimum embedding dimension $d_0 + 1$ is determined when $E(d)$ stops changing for any d_0.

6.3.2 Results for wind speed forecasting

In this section, two application examples to validate the proposed PSO-EKF learning algorithm are presented. First, the experimental analysis of the proposed method applied to the problem of predicting the wind speed is discussed in order to compare the performance with similar approaches [7]. As a second test for the proposed method, the Neural Predictor with experimental data obtained for a microgrid [4] is implemented in order to evaluate the performance with time series of different nature but related with the energy production and demand in smart grids.

6.3.2.1 Comparison of the PSO algorithm for wind speed forecasting

In order to evaluate the performance of the PSO algorithm and compare with similar methods, a neural network predictor for wind speed is im-

plemented on the basis of the proposed training algorithm. The proposed
algorithm requires the following methodology:

1. Define the training set. Training is performed using minute data from
   the first 3 hours from January 1, 2011 and the testing is performed
   using the 3 hours subsequent to the data training (experimental data is
   taken from [15]);
2. Determine the optimal dimension of the regression vector (6.16) for
   data set of step 1;
3. Select the neural structure to be used (6.17);
4. Train the neural identifier and
5. Validate the neural identifier using the testing data.

The neural network used is an RMLP trained with a PSO-EKF, whose
structure is as presented in Fig. 6.12. The hidden layer has 5 units with
logistic sigmoid activation functions (6.18), whose β is fixed in 1 and the
output layer is composed of just one neuron, with a linear activation function. The initial values for the covariance matrices (R, Q, P) are determined
using the PSO algorithm, with 200 as the maximum number of iterations,
4 generations, 3 particles, and $c_1 = c_2 = 0.1$. The initial values for neural
weights are randomly selected. The length of the regression vector is 5 because that is the order of the system, which is determined using the Cao
methodology.

The training is performed off-line using a series–parallel configuration.
For this purpose, the delayed output is taken from the wind speed. The
mean square error (MSE) reached in training is 1.735×10^{-5} in 200 iterations; the mean absolute relative error reached is 0.6912%. Besides, to
measure the performance of the neural network, the absolute relative error
(ARE) is calculated from

$$ARE = \left| \frac{y(n) - \widehat{y}(n)}{y(n)} \right|, \qquad (6.20)$$

where $\widehat{y}(n)$ is the predicted wind speed time series achieved by the network. Simulation results are presented as follows: Fig. 6.13 displays the
wind speed time series neural identification, Fig. 6.14 includes the identification error (6.11), Fig. 6.15 shows the time evolution of mean square
error, and Fig. 6.16 displays the absolute relative error obtained with (6.20).

Fig. 6.17 depicts the comparison detail of the proposed PSO-EKF
training algorithm against that of the classical EKF and the well-known
Levenberg–Marquardt algorithm.

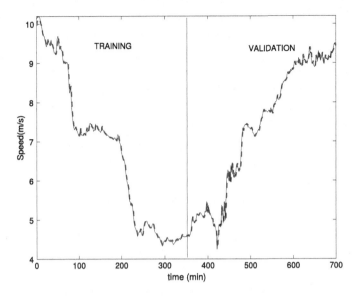

Figure 6.13 Identification performance for wind speed forecast.

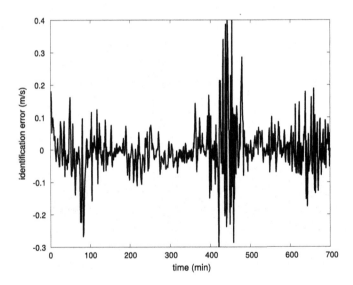

Figure 6.14 Identification error.

Table 6.2 includes a comparison between the proposed PSO–EKF learning algorithm, the EKF and the well-known Levenberg–Marquardt (LM) algorithm.

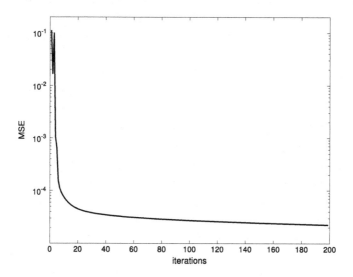

Figure 6.15 Mean square error.

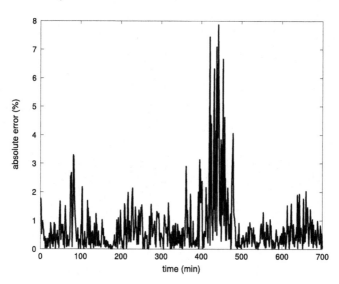

Figure 6.16 Absolute relative error.

Table 6.2 Mean value (μ) and standard deviation (σ) for identification error

	PSO-EKF	EKF	LM
μ	-5.6693×10^{-6}	-1.2547×10^{-5}	1.8354×10^{-4}
σ	0.0749	0.0807	0.0724

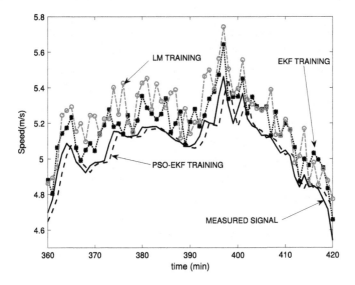

Figure 6.17 Identification performance comparison for wind speed forecast.

Results included in Table 6.2 show that the proposed methodology leads to an improvement of the results. Therefore, for the second example, only PSO-EKF results are presented.

6.3.3 Results for electricity price forecasting

The application of neural networks for the prediction of time series in electric power systems has a great impact in the planning of the markets of this commodity even when other techniques are used to realize the prediction of time series [14], [35]. The results shown in this work represent an important alternative. The purpose of predictions is to reduce risks in decision making, through estimates of future events. Prediction in general has applications in several areas. For the case considered in this section, the prediction of time series in electric power systems has direct application in the planning of generation and supply of electric power and direct influence on the electricity markets. The electric power industry, in many countries, is becoming more competitive day by day. This fact explains the importance of knowing in advance the behavior of the market in order to provide a service of better quality and lower cost for both the producer and the consumer [48]. The proposed methodology described above will be used for electricity price forecasting. Some power exchanges and data vendors openly provide high-frequency (hourly, half-hourly) time series of electric-

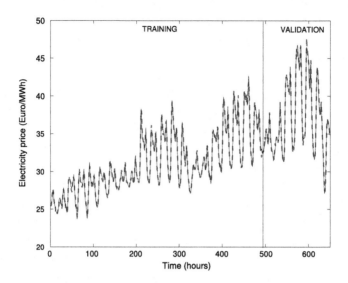

Figure 6.18 Identification performance for electricity prices forecast.

ity prices on their web pages. For instance, Nord Pool publishes price and other fundamental power market data for the most recent two-year period (www.nordpoolspot.com). The following test has been developed with this data for November 2016 following the same procedure as explained above.

The neural network used is an RMLP trained with a PSO-EKF, whose structure is as the presented in Fig. 6.12; the hidden layer has 10 units with logistic sigmoid activation functions (6.18), whose β is fixed in 1 and the output layer is composed of just one neuron, with a linear activation function. The initial values for the covariance matrices (R, Q, P) are determined using the PSO algorithm, with 200 as the maximum number of iterations, 5 generations, 10 particles, and $c_1 = c_2 = 0.1$. The initial values for neural weights are randomly selected. The length of the regression vector is 8 because that is the order of the system, which is determined using the Cao methodology.

The training is performed off-line, using a series–parallel configuration; for this case, the delayed output is taken from the electricity price. The mean square error (MSE) reached in training is 5×10^{-5} in 200 iterations. Simulation results are presented as follows: Fig. 6.18 displays the wind speed time series neural identification, Fig. 6.19 includes the identification error (6.11), Fig. 6.20 shows the time evolution of mean square error, and Fig. 6.21 displays the absolute relative error obtained with (6.20).

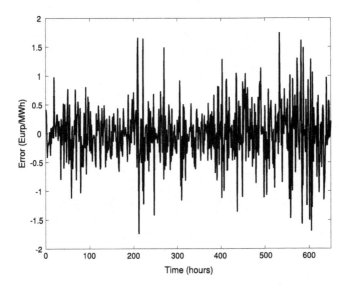

Figure 6.19 Identification error for electricity prices forecast.

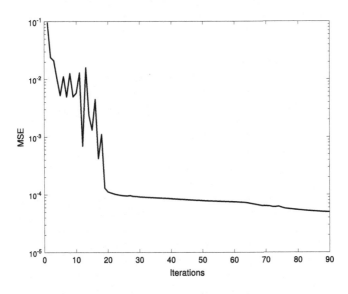

Figure 6.20 Mean square error performance for electricity prices forecast.

6.4. CONCLUSIONS

This chapter has presented the application of recurrent high order neural networks to the identification of discrete-time nonlinear systems.

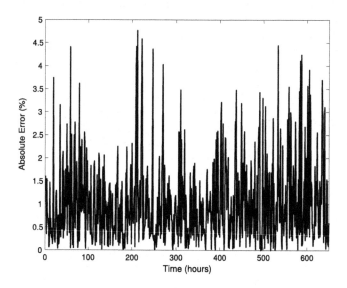

Figure 6.21 Absolute relative error for electricity prices forecast.

The training of the neural networks was performed on-line using an EKF improved with a PSO algorithm. Experimental results illustrate the applicability of the proposed identification methodology for the on-line identification of a three-phase induction motor. In our experiments, the proposed neural identifier proves to be a model that captures very well the complexity for unknown discrete-time nonlinear systems. The use of PSO to improve the identification results has been experimentally illustrated in this chapter. This chapter proposed the use of an RMLP trained with a PSO-EKF learning algorithm, to accurately predict wind speed with good results as shown in Table 6.2. The same methodology was then applied for electricity price forecasting.

The proposed method has a compact structure but takes into account the dynamic nature of the system whose behavior it is required to predict. The proposed neural identifier in our experiments proves to be a model that captures very well the complexity associated with important variables in smart grids operation. Future work on implementing higher order neural networks aims for the design of optimal operation algorithms for smart grids composed of wind and photovoltaic generation systems interconnected to the utility grid. This management system can use the forecasting data to operate the global system, that is, fulfilling the load demand, minimizing the power supplied by the utility grid, and maximizing the one supplied by renewable sources.

REFERENCES

[1] Al-kazemi B, Mohan CK. Multi-phase generalization of the particle swarm optimization algorithm. In: Evolutionary computation, 2002. CEC '02. Proceedings of the 2002 congress on, vol. 1; May 2002. p. 489–94.

[2] Alanis AY, Rangel E, Rivera J, Arana-Daniel N, Lopez-Franco C. Particle swarm based approach of a real-time discrete neural identifier for linear induction motors. Math Probl Eng 2013;2013:9.

[3] Alanis AY, Ricalde LJ, Sanchez EN. High order neural networks for wind speed time series prediction. In: 2009 international joint conference on neural networks; June 2009. p. 76–80.

[4] Alanis AY, Ricalde LJ, Simetti C, Odone F. Neural model with particle swarm optimization Kalman learning for forecasting in smart grids. Math Probl Eng 2013;2013:9.

[5] Alanis AY, Sanchez EN, Loukianov AG. Discrete-time adaptive backstepping nonlinear control via high-order neural networks. IEEE Trans Neural Netw July 2007;18(4):1185–95.

[6] Alanis AY, Sanchez EN, Loukianov AG, Perez-Cisneros MA. Real-time discrete neural block control using sliding modes for electric induction motors. IEEE Trans Control Syst Technol Jan 2010;18(1):11–21.

[7] Alanis AY, Simetti C, Ricalde LJ, Odone F. A wind speed neural model with particle swarm optimization Kalman learning. In: World automation congress 2012; June 2012. p. 1–5.

[8] Benitez VH, Loukianov AG, Sanchez EN. Neural identification and control of a linear induction motor using an $\alpha - \beta$ model. In: Proceedings of the 2003 American control conference, 2003, vol. 5; June 2003. p. 4041–6.

[9] Boldea I, Nasar S. Linear electric actuators and generators. Cambridge University Press; 1997. [Online]. Available from: https://books.google.com.mx/books?id=m6tdywAACAAJ.

[10] Brown R, Hwang P. Introduction to random signals and applied Kalman filtering. v. 1. J. Wiley; 1992.

[11] Cao L. Practical method for determining the minimum embedding dimension of a scalar time series. Physica D 1997;110(1):43–50.

[12] Chandra R, Frean M, Zhang M. Adapting modularity during learning in cooperative co-evolutionary recurrent neural networks. Soft Comput 2012;16(6):1009–20.

[13] Clerc M. The swarm and the queen: towards a deterministic and adaptive particle swarm optimization. In: Proceedings of the 1999 congress on evolutionary computation—CEC99 (Cat. No. 99TH8406), vol. 3; 1999. p. 1957.

[14] Cotter NE. The Stone–Weierstrass theorem and its application to neural networks. IEEE Trans Neural Netw Dec 1990;1(4):290–5.

[15] US Department of energy. National renewable energy laboratory, NREL. [Online]. Available from: https://www.nrel.gov/gis/data-wind.html.

[16] El-Fouly THM, El-Saadany EF, Salama MMA. One day ahead prediction of wind speed and direction. IEEE Trans Energy Convers March 2008;23(1):191–201.

[17] Farrell J, Polycarpou M. Adaptive approximation based control: unifying neural, fuzzy and traditional adaptive approximation approaches. Adaptive and Cognitive Dynamic Systems: Signal Processing, Learning, Communications and Control. Wiley; 2006.

[18] Feldkamp LA, Prokhorov DV, Feldkamp TM. Simple and conditioned adaptive behavior from Kalman filter trained recurrent networks. Neural Netw 2003;16(5–6):683–9.

[19] Ge SS, Zhang J, Lee TH. Adaptive neural network control for a class of MIMO nonlinear systems with disturbances in discrete-time. IEEE Trans Syst Man Cybern, Part B, Cybern Aug 2004;34(4):1630–45.

[20] Gieras J. Linear induction drives. Monographs in Electrical and Electronic Engineering. Clarendon Press; 1994. [Online]. Available from: https://books.google.com.mx/books?id=UOdSAAAAMAAJ.

[21] Haykin S. Kalman filtering and neural networks. Adaptive and Cognitive Dynamic Systems: Signal Processing, Learning, Communications and Control. Wiley; 2001.

[22] Hernandez-Gonzalez M, Sanchez EN, Loukianov AG. Discrete-time neural network control for a linear induction motor. In: 2008 IEEE international symposium on intelligent control; Sept 2008. p. 1314–9.

[23] Jie W. Control technologies in distributed generation system based on renewable energy. In: 2009 3rd international conference on power electronics systems and applications (PESA); May 2009. p. 1–14.

[24] Kariniotakis GN, Stavrakakis GS, Nogaret EF. Wind power forecasting using advanced neural networks models. IEEE Trans Energy Convers Dec 1996;11(4):762–7.

[25] Kazantzis N, Kravaris C. Time-discretization of nonlinear control systems via Taylor methods. Comput Chem Eng 1999;23(6):763–84.

[26] Kennel M, Brown R, Abarbanel HD. Determining the embedding dimension for phase-space reconstruction using a geometrical construction. Phys Rev A 1992;45(6).

[27] Kiran R, Jetti SR, Venayagamoorthy GK. Online training of a generalized neuron with particle swarm optimization. In: The 2006 IEEE international joint conference on neural network proceedings; 2006. p. 5088–95.

[28] Kosmatopoulos EB, Polycarpou MM, Christodoulou MA, Ioannou PA. High-order neural network structures for identification of dynamical systems. IEEE Trans Neural Netw Mar 1995;6(2):422–31.

[29] Leung C-S, Chan L-W. Dual extended Kalman filtering in recurrent neural networks. Neural Netw 2003;16(2):223–39.

[30] Li IH, Wang WY, Su SF, Lee YS. A merged fuzzy neural network and its applications in battery state-of-charge estimation. IEEE Trans Energy Convers Sept 2007;22(3):697–708.

[31] Lin CJ, Chen CH, Lin CT. A hybrid of cooperative particle swarm optimization and cultural algorithm for neural fuzzy networks and its prediction applications. IEEE Trans Syst Man Cybern, Part C, Appl Rev Jan 2009;39(1):55–68.

[32] Lin C-J, Tsai H-M. {FPGA} implementation of a wavelet neural network with particle swarm optimization learning. Math Comput Model 2008;47(9–10):982–96.

[33] Loukianov AG, Rivera J, Cañedo JM. Discrete-time sliding mode control of an induction motor. In: IFAC proceedings volumes, vol. 35, no. 1; 2002. p. 19–24.

[34] Meiqin M, Ming D, Jianhui S, Chang L, Min S, Guorong Z. Testbed for microgrid with multi-energy generators. In: 2008 Canadian conference on electrical and computer engineering; May 2008. p. 637–40.

[35] Nogales FJ, Contreras J, Conejo AJ, Espinola R. Forecasting next-day electricity prices by time series models. IEEE Trans Power Syst May 2002;17(2):342–8.

[36] Norgaard M, Poulsen NK, Ravn O. Advances in derivative-free state estimation for nonlinear systems. Technical University of Denmark; 2000.

[37] Parsopoulos K. Particle swarm optimization and intelligence: advances and applications: advances and applications. Advances in Computational Intelligence and Robotics. Information Science Reference; 2010.

[38] Poznyak A, Sanchez E, Yu W. Differential neural networks for robust nonlinear control: identification, state estimation and trajectory tracking. World Scientific; 2001.

[39] Ricalde LJ, Catzin GA, Alanis AY, Sanchez EN. Higher order wavelet neural networks with Kalman learning for wind speed forecasting. In: 2011 IEEE symposium on computational intelligence applications in smart grid (CIASG); April 2011. p. 1–6.

[40] Ricalde LJ, Sanchez EN. Inverse optimal nonlinear recurrent high order neural observer. In: Proceedings. 2005 IEEE international joint conference on neural networks, 2005, vol. 1; July 2005. p. 361–5.

[41] Richert D, Masaud K, Macnab CJB. Discrete-time weight updates in neural-adaptive control. Soft Comput 2013;17(3):431–44.

[42] Rovithakis G, Christodoulou M. Adaptive control with recurrent high-order neural networks: theory and industrial applications. Advances in Industrial Control. Springer; 2000.

[43] Senjyu T, Yona A, Urasaki N, Funabashi T. Application of recurrent neural network to long-term-ahead generating power forecasting for wind power generator. In: 2006 IEEE PES power systems conference and exposition; Oct 2006. p. 1260–5.

[44] Shi Y, Eberhart R. A modified particle swarm optimizer. In: 1998 IEEE international conference on evolutionary computation proceedings. IEEE world congress on computational intelligence (Cat. No. 98TH8360); May 1998. p. 69–73.

[45] Su T, Jhang J, Hou C. Neural networks and particle swarm optimization for function approximation. Int J Innov Comput Inf Control Sept 2008:2363–74.

[46] Takahashi I, Ide Y. Decoupling control of thrust and attractive force of a LIM using a space vector control inverter. IEEE Trans Ind Appl Jan 1993;29(1):161–7.

[47] Welch RL, Ruffing SM, Venayagamoorthy GK. Comparison of feedforward and feedback neural network architectures for short term wind speed prediction. In: 2009 international joint conference on neural networks; June 2009. p. 3335–40.

[48] Weron R. Electricity price forecasting: a review of the state-of-the-art with a look into the future. Int J Forecast 2014;30(4):1030–81.

[49] Williams RJ, Zipser D. A learning algorithm for continually running fully recurrent neural networks. Neural Comput June 1989;1(2):270–80.

[50] Wu YK, Hong JS. A literature review of wind forecasting technology in the world. In: 2007 IEEE Lausanne power tech; July 2007. p. 504–9.

[51] Yu W. Nonlinear system identification using discrete-time recurrent neural networks with stable learning algorithms. Inf Sci 2004;158:131–47.

[52] Zahiri S-H, Seyedin S-A. Swarm intelligence based classifiers. J Franklin Inst 2007;344(5):362–76.

CHAPTER SEVEN

Bio-inspired Algorithms to Improve Neural Controllers for Discrete-time Unknown Nonlinear System

Contents

7.1. NEURAL SECOND-ORDER SLIDING MODE CONTROLLER FOR UNKNOWN DISCRETE-TIME NONLINEAR SYSTEMS

To cope with various search problems in complex systems of the real world, scientists have been inspired by nature for years, both as a model and as a metaphor. Optimization is the core phenomenon of many natural processes and the behavior of social groups of insects, birds, and foraging strategy of microbial creatures. Natural selection tends to eliminate the species with poor feeding strategies to promote the spread of genes of the species with successful foraging behavior. Since during foraging the body

Bio-inspired Algorithms for Engineering
https://doi.org/10.1016/B978-0-12-813788-8.00007-X
© 2018 Elsevier Inc.
All rights reserved.
107

an animal takes the necessary steps to maximize the energy used per unit of time—taking into account all the limitations of their own physiology as detection and cognitive abilities, the environment, the search strategy of natural food—can lead to optimization, essentially, this idea can be applied to optimization problems of other domains of the real world [1], [2], [3].

Recently, several swarm intelligence algorithms have been proposed, such as Ant Colony Optimization (ACO) [4], Particle Swarm Optimization (PSO) [5], [6], [7], [8], Artificial bee colony (ABC) [9], and Bacterial Foraging Optimization (BFO) [10]. The BFO algorithm was first proposed in 2002 by Passino. It is inspired by the foraging behavior and chemotactic bacteria, especially *Escherichia coli* (*E. coli*) in our intestine. By running smooth and tumbling, it can be moved to the escape area of nutrients and poison zone in the environment. Chemotaxis is more attractive behavior of bacteria, and it has been studied by many researchers [1], [11], [12].

Due to their group response, the social behavior of the colony of *E. coli* is very interesting for engineering since it allows them to get quickly and easily the best food supply with the lowest possible risk. These bacteria can communicate through chemical exchanges. The bacteria that have achieved a safe place to feed, communicate it to others who then come to such place; the more the food, the stronger is the signal. Similarly, if the bacteria are in a dangerous place, with agents that may threaten the colony, they warn others to stay away from that place. This foraging behavior can be represented mathematically as a kind of swarm intelligence [1]. It has attracted significant attention from researchers and has been used in several areas of control application, such as design of multiple optimal power system stabilizers [13] and optimization of active power filter for load compensation [14]. It has been shown that BFO is able to find the optimum value and avoid being trapped in the local optima. In the optimization process, besides its ability to locate the optimum value, the convergence speed of BFO is also considered [15].

On the other hand, the PSO technique, which is based on the behavior of a flock of birds or school of fish, is a type of evolutionary computing technique [16]. The development of particle swarm optimization is based on concepts and rules that govern socially organized populations in nature, such as bird flocks, fish schools, and animal herds. Unlike the ant colony approach, where stigmergy is the main communication mechanism among individuals through their environment, in such systems communication is rather direct without altering the environment [17]. Since its inception in 1995, research and application interest in PSO have increased, resulting in an exponential increase in the number of publications; applications vary in complexity and cover a wide range of application areas [17]. Therefore, the main contribution of this chapter is the use of PSO to improve a neural second-order sliding mode controller for unknown discrete-time nonlinear systems, by adapting control and learning parameters instead of using heuristic solutions.

Otherwise, different approaches can be taken for real-time implementations of control systems using digital devices. One is to obtain a continuous-time system model, then synthesize a continuous-time controller using well-established methodologies, and thereafter discretize it. Another way is to obtain a discrete-time system model directly by identification or by integration in time of a continuous-time model and then synthesize a discrete-time controller, which can be implemented digitally [18]. Sliding mode control is principally characterized by its robustness with respect to the system's modeling uncertainties and external disturbances. However, the application of this kind of control law is confronted by a serious problem of chattering. Owing to the many advantages of the digital control strategy, the discretization of the sliding mode control (SMC) has become an interesting research field. Unfortunately, the chattering phenomenon is more obvious in this case, because the sampling rate is smaller [19], [20]. In this way, many approaches have been suggested in order to solve this problem. In the eighties, a new control technique called high-order sliding mode control was investigated. Its main idea is to reduce to zero not only the sliding function, but also its high-order derivatives [20]. This approach allows to reduce the oscillations amplitude, besides, the outstanding sliding mode systems robustness remains intact [21].

However, although this kind of control is robust to external disturbances [20], it requires previous knowledge of a nominal system model. To overcome this requirement—in this work—we propose the use of a neural identifier that provides a mathematical model for the system to be controlled. In this connection, it is well known that Artificial Neural Networks (ANN) exhibit interesting properties such as adaptability, learning capability, and ability to generalize. Due to these properties, ANN have been established as an excellent methodology as exemplified by their applications to identification and control of nonlinear and complex systems [22]. However, training ANN is a complex task of great importance in the supervised learning topic particularly for real-life complex problems, which require the integration of several of Soft Computing methodologies to achieve the efficiency and accuracy needed in practice [23], [3], [24], [7],

[25]. The best well-known training approach for recurrent neural networks (RNN) is the back propagation through time learning [26]. However, it is a first-order gradient descent method, and hence, its learning speed could be very slow [26].

In past years the EKF-based algorithms have been introduced to train neural networks in order to improve the learning convergence [26]. Besides, it has proven to be reliable and practical for many applications [26]. However, EKF training usually requires the heuristic selection of some design parameters, which is not always an easy task [18], [5], [27].

Therefore, in this chapter various nonlinear techniques are combined in order to develop a second-order sliding mode controller for discrete-time unknown MIMO nonlinear systems which can include both external and internal disturbances and does not require previous knowledge of a nominal model. To achieve this goal, it is necessary to combine high-order sliding mode technique with neural identification and an on-line learning algorithm based on EKF. This controller requires a suitable selection of design parameters for control law, as well as for learning algorithm. Typically, such parameters are heuristically selected. However, in this chapter an approach based on BFO is proposed to do an automatic selection of such parameters, resulting in a significant improvement for the controller development, thereby combining the above-mentioned nonlinear techniques with a bio-inspired optimization one.

7.1.1 Second-Order Sliding Mode Controller

HOSM is actually a movement on a discontinuity set of a dynamic system understood in Filippov's sense [28]. The sliding order characterizes the dynamics smoothness degree in the vicinity of the mode. If the task is to provide for keeping a constraint given by equality of a smooth function to zero, the sliding order is a number of continuous total derivatives of (including the zero one) in the vicinity of the sliding mode [29]. Hence, the rth-order sliding mode is determined by the equalities

$$S(t, x) = \dot{S}(t, x) = \ddot{S}(t, x) = \cdots = S(t, x)^{(r-1)} = 0, \qquad (7.1)$$

forming an r-dimensional condition on the state of the dynamic system. The words "rth-order sliding" are often abridged to "r-sliding". In this notation, standard sliding mode is called 1-sliding mode since $\dot{S}$ is discontinuous [29], [30]. Higher-order sliding modes (HOSM) preserve the features of the first-order sliding modes and can improve them, if properly designed. The latter is done by eliminating the chattering [31]. The aim of the first-order sliding mode control is to force the state to move on the switching surface $S(t, x)$. In high-order sliding mode control, the purpose is to force the state to move on the switching surface $S(t, x)$ and to keep its $(r - 1)$ first successive derivatives null [19], [20]. For the second-order sliding mode control, the following relation must be verified:

$$S(t, x) = \dot{S}(t, x) = 0. \tag{7.2}$$

Among second-order sliding mode control algorithms, Bartolini's approach and some similar approaches (twisting and supertwisting [21]) are highly appreciated as they do not require knowledge of the system parameters. Yet, if some information, even uncertain, about the system is accessible, it would be favorable to exploit it in order to ameliorate the closed loop system's performance. In fact, the equivalent control approach for second-order sliding mode requires the use of a system's model and allows an asymptotic convergence of the sliding function to zero [19], [20]. In the following paragraph, a second-order sliding mode control with asymptotic sliding function convergence is proposed.

Let us consider the system defined by

$$\begin{aligned} x(k+1) &= F(x(k)) + Bu(k), \\ y(k) &= Hx(k), \end{aligned} \tag{7.3}$$

where $x \in \Re^n$, $u \in \Re^m$, $y \in \Re^p$, $F \in \Re^n \times \Re^m \to \Re^n$, $B \in \Re^{n \times m}$, and $H \in \Re^{p \times n}$. The sliding function is selected as [20]

$$S(k) = C^{\mathsf{T}}(x(k) - x_d(k)) \tag{7.4}$$

with $x_d(k)$ as the desired state vector. The new sliding function $\sigma(k)$ is defined by

$$\sigma(k) = S(k+1) + \beta S(k) \tag{7.5}$$

with $\beta \in [0, 1[$ to ensure the stability of $\sigma(k)$ [20], and

$$\begin{aligned} S(k+1) &= C^{\mathsf{T}}(x(k+1) - x_d(k+1)) \\ &= C^{\mathsf{T}}(F(x(k)) + Bu(k) - x_d(k+1)). \end{aligned}$$

The equivalent control that forces the system to evolve on the sliding surface is deduced from

$$\sigma(k+1) = \sigma(k) = 0.$$

Then

$$S(k+1) + \beta S(k) = 0$$

and

$$
\begin{aligned}
S(k+1) &= \sigma(k+1) - \beta S(k) \\
&= C^{\mathsf{T}}(x(k+1) - x_d(k+1)) \\
&= C^{\mathsf{T}}(F(x(k)) + Bu(k) - x_d(k+1));
\end{aligned}
$$

therefore

$$u_{eq}(k) = (C^{\mathsf{T}}B)^{-1}\left[-\beta S(k) - C^{\mathsf{T}}F(x(k)) + C^{\mathsf{T}}x_d(k+1)\right]. \tag{7.6}$$

The robustness is ensured by the addition of a discontinuous term (sign of the new sliding function $\sigma(k)$). By analogy with the continuous-time case, the integral of the discontinuous term, approximated by a first-order transformation [20], is applied to system (7.3):

$$u_{dis}(k) = u_{dis}(k-1) - TMsign(\sigma(k)) \tag{7.7}$$

with T as the sampling step. The control at the instant k is then

$$u(k) = u_{eq}(k) + u_{dis}(k). \tag{7.8}$$

7.1.2 Neural Second-Order Sliding Mode Controller

In order to design a neural second-order sliding mode controller, it is necessary to consider the Separation Principle for MIMO discrete-time nonlinear systems [32] as follows.

Theorem 1 (Separation Principle). *[32] The asymptotic stabilization problem of an MIMO nonlinear system*

$$
\begin{aligned}
\chi(k+1) &= F(\chi(k), u(k)), &\tag{7.9} \\
y(k) &= h(x(k)), &\tag{7.10}
\end{aligned}
$$

where $\chi \in \Re^n$, $u \in \Re^m$, and $F \in \Re^n \times \Re^m \to \Re^n$ is a nonlinear function, via estimated state feedback

$$
\begin{aligned}
u(k) &= \xi(\widehat{\chi}(k)), \\
\widehat{\chi}(k+1) &= F(\widehat{\chi}(k), u(k), y(k)) \tag{7.11}
\end{aligned}
$$

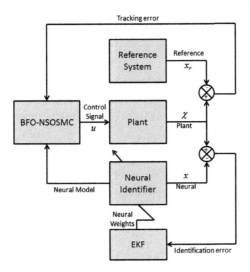

Figure 7.1 Neural second-order sliding mode control scheme.

is solvable if and only if system (7.9)–(7.10) is asymptotically stabilizable and exponentially detectable.

Corollary 1. *[32] There is an exponential observer for a Lyapunov stable discrete-time nonlinear system (7.9)–(7.10) with u = 0 if and only if the linear approximation*

$$\chi\left(k+1\right) = A\left(k\right)\chi\left(k\right) + Bu\left(k\right),$$
$$y\left(k\right) = C\chi\left(k\right), \tag{7.12}$$
$$A = \left.\frac{\partial F}{\partial \chi}\right|_{\chi=0}, \quad B = \left.\frac{\partial F}{\partial u}\right|_{\chi=0}, \quad C = \left.\frac{\partial h}{\partial x}\right|_{\chi=0}$$

of system (7.9)–(7.10) is detectable.

Now, it is possible to design a neural second-order sliding mode controller as is shown in Fig. 7.1. Its validity is based on the following proposition [18], [5], [27].

Proposition 1. *[5] Given a desired output trajectory $y_r = x_1^d$, a dynamic system with output y, and a neural network with output y_n, it is possible to establish the following inequality:*

$$\left\|y_r - y\right\| \leq \left\|y_n - y\right\| + \left\|y_r - y_n\right\|,$$

where $y_r - y$ is the system output tracking error, $y_n - y$ is the output identification error, and $y_r - y_n$ is the RHONN output tracking error.

This proposition is possible based on the separation principle for discrete-time nonlinear systems [32] as stated in Theorem 1 and Corollary 1 with $C = I$.

Based on Proposition 1, it is possible to divide the tracking error in two parts:

1. Minimization of $y_n - y$, which can be achieved by the proposed on-line identification algorithm as explained in Section 6.2.1 of Chapter 6;

2. Minimization of $y_n - y_r$. To do so, a tracking algorithm is developed on the basis of the neural model. This can be reached by designing a control law based on the RHONN model. To design such controller, we propose using the discrete-time second-order sliding mode methodology [20] as established in the previous section [18], [5], [27].

7.1.3 Simulation results of the Neural Second-Order Sliding Mode Controller

In order to illustrate the applicability of the proposed scheme, in this section, the scheme is applied to a forced Van der Pol oscillator, whose discretized nonlinear dynamics is represented by the following equation system [33]:

$$
\begin{aligned}
\chi_1(k+1) &= \chi_1(k) + T\chi_2(k), \\
\chi_2(k+1) &= \chi_2(k) + T\left(\xi\left(0.5 - x_1^2(k)\right)x_2(k)\right) \\
&\quad + T\left(-\chi_1(k) + u(k)\right), \\
y(k) &= \chi_1(k), \\
u(k) &= 0.5\cos(1.1kT),
\end{aligned}
\tag{7.13}
$$

where variables $x \in \Re^2$, $u \in \Re$, and $y \in \Re$ are the state, input, and output of the system, respectively, and T is the sampling time, which is fixed at 1×10^{-3} s, and ξ is a parameter whose nominal value is 1. However, for this application, such a parameter is used to include a parametric variation (which is assumed unknown).

To identify the model (7.13), we use the RHONN (7.14) with $n = 2$ trained with the EKF, as explained in Chapter 6:

$$
\begin{aligned}
x_1(k+1) &= w_{11}(k)S(\chi_1(k)) + w_{12}(k)S^2(\chi_2(k)), \\
x_2(k+1) &= w_{21}(k)S(\chi_1(k)) + w_{22}(k)S^2(\chi_2(k)) \\
&\quad + w_{23}u(k).
\end{aligned}
\tag{7.14}
$$

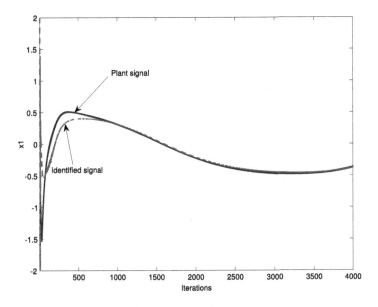

Figure 7.2 Identification for x_1.

The training is performed on-line, using a parallel configuration as displayed in Fig. 7.1. Both the NN and Van der Pol system states are initialized randomly. The associated covariance matrices are initialized as diagonals, and the nonzero elements are $P_1(0) = P_2(0) = 100000$; $Q_1(0) = Q_2(0) = 1000$, and $R_1(0) = R_2(0) = 10000$, respectively. The control law is computed with (7.8) for the RHONN model (7.14). The simulation results are presented in Figs. 7.2 and 7.3; they display the time evolution of the identified states $x_1(k)$ and $x_2(k)$, respectively. Fig. 7.4 shows the identification errors. The weights evolution is included in Fig. 7.5. Finally, the tracking performance is presented in Fig. 7.6.

7.2. NEURAL-PSO SECOND-ORDER SLIDING MODE CONTROLLER FOR UNKNOWN DISCRETE-TIME NONLINEAR SYSTEMS

This section deals with adaptive tracking for unknown discrete-time MIMO nonlinear systems in the presence of disturbances. A PSO is used to improve a discrete-time neural second-order sliding mode controller for unknown discrete-time nonlinear systems [18], [5], [27]. In order to show

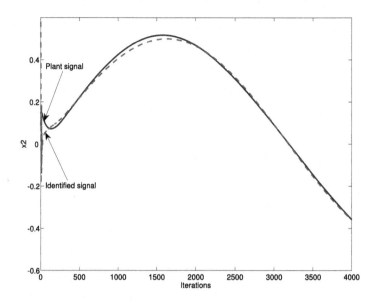

Figure 7.3 Identification of x_2.

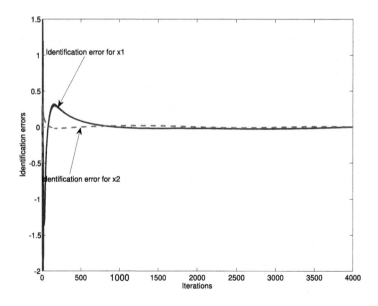

Figure 7.4 Identification errors.

the applicability of the proposed scheme, simulation results are included for a Van der Pol oscillator.

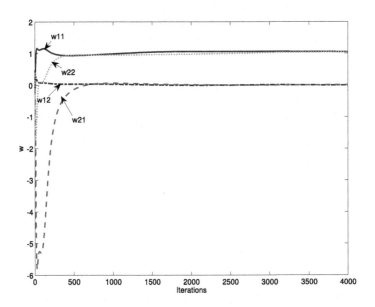

Figure 7.5 Time evolution for adaptive weights.

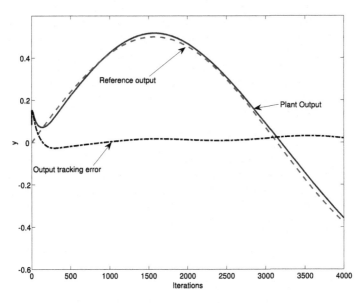

Figure 7.6 Tracking output (solid line), tracking error (dash-dot line) and reference signal (dash line).

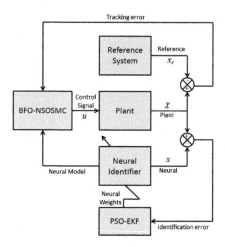

Figure 7.7 Neural-PSO second-order sliding mode control scheme.

7.2.1 Neural-PSO Second-Order Sliding Mode Controller design

From (7.7) it is easy to see that success of the proposed controller depends on the selection of gain M. Due to the PSO benefits explained above, in this chapter, PSO was used to optimize the discrete–time high–order sliding mode control law (7.7) by adapting the control gain M instead of heuristic solutions. PSO was also used to improve the EKF–based training algorithm. The proposed control scheme is depicted in Fig. 7.7 [18], [5], [27].

7.2.2 Simulation results of the Neural-PSO Second-Order Sliding Mode Controller

The training is performed on–line using a parallel configuration as displayed in Fig. 7.7. Both the NN and Van der Pol system states are initialized randomly. The associated covariance matrices are computed using the PSO algorithm, and the RHONN weights are updated with EKF as explained in Chapter 6. The control law is computed with (7.8) for the RHONN model (7.14), with an adapting control gain M computed with the PSO algorithm.

The simulation results are presented as follows:

1. Fig. 7.8 displays the time evolution of the output trajectory tracking,
2. Fig. 7.9 shows the computed control signal, and
3. The output tracking error is presented in Fig. 7.10.

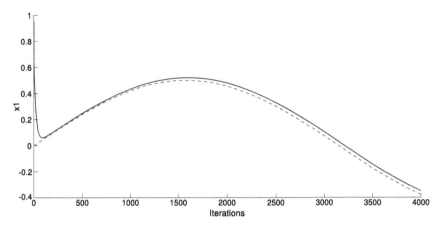

Figure 7.8 Output trajectory tracking (plant signal is in dash line and reference in solid line).

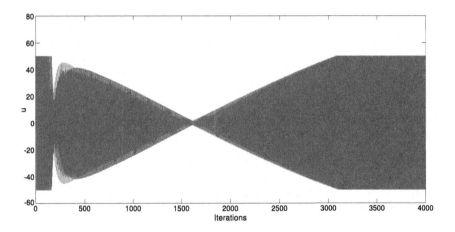

Figure 7.9 Applied control signal.

7.3. NEURAL-BFO SECOND-ORDER SLIDING MODE CONTROLLER FOR UNKNOWN DISCRETE-TIME NONLINEAR SYSTEMS

In this section, various nonlinear techniques are combined in order to develop a second-order sliding mode controller for discrete-time unknown MIMO nonlinear systems, which can include both external and internal disturbances and does not require previous knowledge of a nominal model. To achieve this goal, it is necessary to combine high-order sliding mode

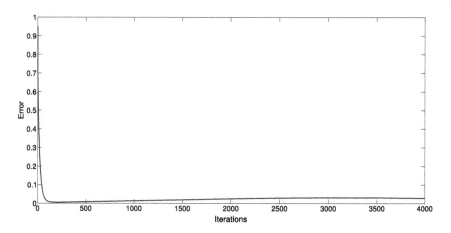

Figure 7.10 Output trajectory tracking error.

technique with neural identification and an on-line learning algorithm based on an EKF. This controller requires a suitable selection of design parameters for control law and learning algorithm. Typically, such parameters are heuristically selected. However, in this chapter, an approach based on BFO is proposed to do an automatic selection of such parameters. The latter resulted in significant controller design improvement. Combining the above-mentioned nonlinear techniques with a bio-inspired optimization technique is illustrated in [18], [5], [27].

7.3.1 Neural-BFO Second-Order Sliding Mode Controller design

From (7.7) it is easy to see that success of the proposed controller depends on the selection of gain M. Due to the BFO benefits as explained before, in this chapter, BFO is used to optimize the discrete-time high-order sliding mode control law (7.7) by adapting the control gain M instead of heuristic solutions. Besides, BFO is also used to improve the EKF-based training algorithm. The proposed control scheme is depicted in Fig. 7.11.

Natural selection tends to eliminate animals with poor foraging strategies (methods for locating, handling, and ingesting food) and promotes the propagation of genes of those animals that have successful foraging strategies. Species who have better food searching ability are capable of enjoying reproductive success, and those with poor searching ability are either eliminated or reshaped.

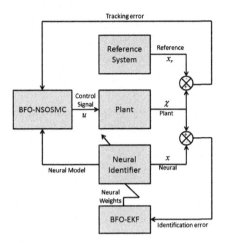

Figure 7.11 Neural-BFO second-order sliding mode control scheme.

The proposed algorithm mimics the foraging behavior of *E. coli* present in our intestines. It is categorized into the following processes: Chemotaxis, Reproduction, and Elimination [13], [34].

(1) *Chemotaxis*: This process simulates the movements of an *E. coli* cell through swimming and tumbling using flagella [13], the number of chemotactic steps N_c, and the direction of movement after a tumble given by

$$\theta^i(j+1, k, l) = \theta(j, k, l) + C(i) \times \varphi(j), \tag{7.15}$$

where $C(i)$ is the step size taken in direction of the tumble, j is the index for the chemotactic step taken, k is the index for the number of reproduction step, l is the index for the number of elimination-dispersal event, and $\varphi(j)$ is the unit length random direction taken at each step. If the cost at $\theta^i(j+1, k, l)$ is better than the cost at $\theta^i(j, k, l)$, then the bacterium takes another step of size $C(i)$ in that direction. This process will be continued while the number of steps taken is not greater than N_s.

(2) *Reproduction*: The health of each bacterium is calculated as the sum of the step fitness during its life:

$$J^i_{health} = \sum_{j=1}^{N_c+1} J\left(i, j, k, l\right). \tag{7.16}$$

All bacteria are sorted in increasing order according to health status. In the reproduction step, only the first half of population survives and each surviving bacterium reproduces by splitting into two daughter bacteria, which are then placed at the same locations. Thus, the population of bacteria remains constant [34].

(3) *Elimination-Dispersal*: Chemotaxis provides a basis for the local search, and reproduction process accelerates the convergence of the algorithm [34]. However, the processes of chemotaxis and reproduction are insufficient for global optimal search. Because bacteria can get stuck around the initial positions or local optima, it is possible to use the elimination-dispersal process to change gradually or suddenly in order to avoid being trapped in a local optimum [34]. Therefore, for each elimination-dispersal event, each bacterium is eliminated with a probability of p_{ed}. A low value of N_{ed} dictates that the algorithm will not rely on random elimination-dispersal events to try to find favorable regions. A high value increases computational complexity but allows bacteria to find favorable regions. The p_{ed} should not be large either or else it would lead to an exhaustive search.

Due the BFO properties—extensively explained in literature [34]—BFO are very suitable models for noisy problems, just as the one we are considering. Besides, BFO has shown good results in solving optimization problems [13]. It will be used to optimize the values of Kalman's filter covariance matrices instead of heuristic solutions. For this purpose, each bacteria will represent one of the Kalman covariance entries.

7.3.2 Simulation results of the Neural-BFO Second-Order Sliding Mode Controller

The training is performed on-line using a parallel configuration as displayed in Fig. 7.11. Both the NN and Van der Pol system states are initialized randomly. The associated covariance matrices are computed using the BFO algorithm (7.15) and (7.16), and the RHONN weights are updated with the EKF. The control law is computed with (7.8) for the RHONN model (7.14), with an adapting control gain M computed with the BFO algorithm (7.15)–(7.16).

The simulation results are presented as follows:

1. Fig. 7.12 displays the time evolution of the output trajectory tracking, and its details are depicted in Fig. 7.13,

2. Fig. 7.14 shows the computed control signal, and

3. The output tracking error is presented in Fig. 7.15.

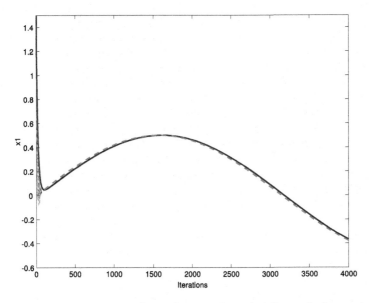

Figure 7.12 Output trajectory tracking (plant signal is in dash line and reference in solid line).

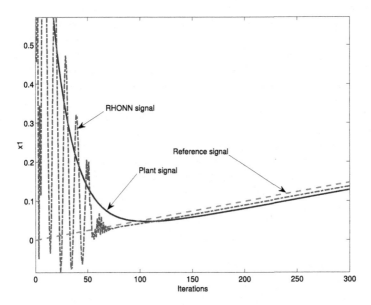

Figure 7.13 Detail for output trajectory tracking (plant signal is in dash line, neural signal in dash-dot line, and reference signal in solid line).

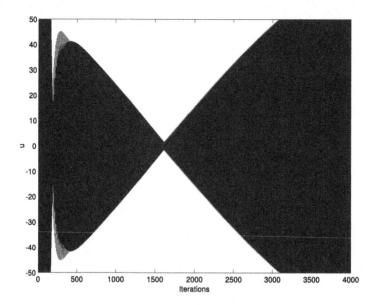

Figure 7.14 Applied control signal.

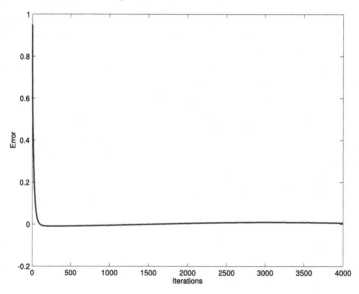

Figure 7.15 Output trajectory tracking error.

7.4. COMPARATIVE ANALYSIS

In order to compare the proposed control scheme with respect to previous works as presented in [18] and [5], Table 7.1 is included. The

Table 7.1 Mean value and standard deviation for tracking errors

Controller	Mean value	Standard deviation
SOSMC	0.0362	0.1114
NSOSMC	0.0073	0.0408
PSO-NSOSMC	0.0043	0.0233
BFO-NSOSMC	0.0037	0.0209

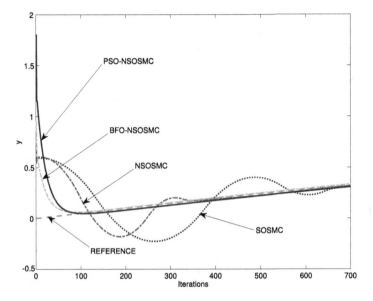

Figure 7.16 Comparative results for output trajectory tracking.

controllers used in this comparison are: 1) Second-Order Sliding Mode Controller (SOSMC) [18], 2) Neural Second-Order Sliding Mode Controller (NSOSMC) [18], 3) PSO gain selection for a Neural Second-Order Sliding Mode Controller (PSO-NSOSMC) [5], and 4) the proposed BFO gain selection for a Neural Second-Order Sliding Mode Controller (BFO-NSOSMC).

It is important to note that the inclusion of BFO for parameter design selection of the proposed Neural Second-Order Sliding Mode Controller significantly improved the performance of the controller even compared with the PSO-NSOSMC. Besides, the BFO-NSOSMC requires neither previous knowledge of the plant parameters nor the plant model. Fig. 7.16 shows the comparative results for output trajectory tracking subject to the above-described controllers.

7.5. CONCLUSIONS

This chapter has presented a discrete-time neural second-order sliding mode controller. PSO and BFO were applied to improve the proposed controller. Simulation results illustrated the effectiveness of the proposed control scheme. Based on results conveyed by Table 7.1 and Fig. 7.6, the BFO proposed scheme has a better performance than similar ones.

REFERENCES

[1] Bhushan Bharat, Singh Madhusudan. Adaptive control of DC motor using bacterial foraging algorithm. Appl Soft Comput December 2011;11(8):4913–20.

[2] Cervantes Leticia, Castillo Oscar. Genetic optimization of membership functions in modular fuzzy controllers for complex problems. Berlin, Heidelberg: Springer Berlin Heidelberg; 2013. p. 51–62.

[3] Cervantes L, Castillo O. Statistical comparison of type-1 and type-2 fuzzy systems design with genetic algorithms in the case of three tank water control. In: 2013 joint IFSA world congress and NAFIPS annual meeting (IFSA/NAFIPS); June 2013. p. 1056–61.

[4] Dorigo M, Gambardella LM. Ant colony system: a cooperative learning approach to the traveling salesman problem. IEEE Trans Evol Comput Apr 1997;1(1):53–66.

[5] Alanis AY, Arana-Daniel N, Lopez-Franco C. Neural-PSO second order sliding mode controller for unknown discrete-time nonlinear systems. In: The 2013 international joint conference on neural networks (IJCNN); Aug 2013. p. 1–6.

[6] Kennedy J, Eberhart R. Particle swarm optimization (PSO). In: Proc. IEEE international conference on neural networks; 1995. p. 1942–8.

[7] Maldonado Yazmin, Castillo Oscar, Melin Patricia. Particle swarm optimization of interval type-2 fuzzy systems for {FPGA} applications. Appl Soft Comput 2013;13(1):496–508.

[8] Valdez Fevrier, Melin Patricia, Castillo Oscar. Modular neural networks architecture optimization with a new nature inspired method using a fuzzy combination of particle swarm optimization and genetic algorithms. Inf Sci 2014;270:143–53.

[9] Karaboga Dervis. An idea based on honey bee swarm for numerical optimization. Technical report-tr06, Erciyes University; 2005.

[10] Passino KM. Biomimicry of bacterial foraging for distributed optimization and control. IEEE Control Syst Jun 2002;22(3):52–67.

[11] Das S, Dasgupta S, Biswas A, Abraham A, Konar A. On stability of the chemotactic dynamics in bacterial-foraging optimization algorithm. IEEE Trans Syst Man Cybern, Part A, Syst Hum May 2009;39(3):670–9.

[12] Dasgupta S, Das S, Abraham A, Biswas A. Adaptive computational chemotaxis in bacterial foraging optimization: an analysis. IEEE Trans Evol Comput Aug 2009;13(4):919–41.

[13] Das TK, Venayagamoorthy GK, Aliyu UO. Bio-inspired algorithms for the design of multiple optimal power system stabilizers: SPPSO and BFA. IEEE Trans Ind Appl Sept 2008;44(5):1445–57.

[14] Mishra S, Bhende CN. Bacterial foraging technique-based optimized active power filter for load compensation. IEEE Trans Power Deliv Jan 2007;22(1):457–65.

[15] Supriyono H, Tokhi MO, Zain BAM. Control of a single-link flexible manipulator using improved bacterial foraging algorithm. In: 2010 IEEE conference on open systems (ICOS 2010); Dec 2010. p. 68–73.

[16] Kiran R, Jetti SR, Venayagamoorthy GK. Online training of a generalized neuron with particle swarm optimization. In: The 2006 IEEE international joint conference on neural network proceedings; 2006. p. 5088–95.

[17] Parsopoulos KE. Particle swarm optimization and intelligence: advances and applications: advances and applications. Advances in Computational Intelligence and Robotics. Information Science Reference; 2010.

[18] Alanis AY, Sanchez EN. Second order sliding mode for MIMO nonlinear uncertain systems based on a neural identifier. In: 2010 world automation congress; Sept 2010. p. 1–6.

[19] Mihoub M, Nouri AS, Abdennour R Ben. An asymptotic discrete second order sliding mode control law for highly non stationary systems. In: 2008 IEEE international conference on emerging technologies and factory automation; Sept 2008. p. 738–44.

[20] Mihoub Mohamed, Nouri Ahmed Said, Abdennour Ridha Ben. Real-time application of discrete second order sliding mode control to a chemical reactor. Control Eng Pract 2009;17(9):1089–95.

[21] Levant Arie. Sliding order and sliding accuracy in sliding mode control. Int J Control 1993;58(6):1247–63.

[22] Ge SS, Zhang Jin, Lee Tong Heng. Adaptive neural network control for a class of MIMO nonlinear systems with disturbances in discrete-time. IEEE Trans Syst Man Cybern, Part B, Cybern Aug 2004;34(4):1630–45.

[23] Konar A. Artificial intelligence and soft computing: behavioral and cognitive modeling of the human brain. CRC Press; 1999.

[24] Castillo O, Melin P. Hybrid intelligent systems for time series prediction using neural networks, fuzzy logic, and fractal theory. IEEE Trans Neural Netw Nov 2002;13(6):1395–408.

[25] Melin Patricia, Astudillo Leslie, Castillo Oscar, Valdez Fevrier, Garcia Mario. Optimal design of type-2 and type-1 fuzzy tracking controllers for autonomous mobile robots under perturbed torques using a new chemical optimization paradigm. Expert Syst Appl 2013;40(8):3185–95.

[26] Singhal Sharad, Wu Lance. In: Advances in neural information processing systems 1; 1989. p. 133–40.

[27] Alanis Alma Y, Arana-Daniel Nancy, Lopez-Franco Carlos. Bacterial foraging optimization algorithm to improve a discrete-time neural second order sliding mode controller. Appl Math Comput 2015;271:43–51.

[28] Arscott FM, Filippov AF. Differential equations with discontinuous righthand sides: control systems. Mathematics and its Applications. Netherlands: Springer; 2013.

[29] Levant A. Introduction to high-order sliding modes. Technical report. School of Mathematical Sciences, Israel. Available from: http://www.tau.ac.il/~levant/hosm2002.pdf, 2003.

[30] Utkin VI. Sliding modes in control and optimization. Communications and Control Engineering. Springer Berlin Heidelberg; 2013.

[31] Moreno JA, Osorio M. A Lyapunov approach to second-order sliding mode controllers and observers. In: 2008 47th IEEE conference on decision and control; Dec 2008. p. 2856–61.

[32] Lin Wei, Byrnes CI. Design of discrete-time nonlinear control systems via smooth feedback. IEEE Trans Autom Control Nov 1994;39(11):2340–6.

[33] Zhu Quanmin, Guo Lingzhong. Stable adaptive neurocontrol for nonlinear discrete-time systems. IEEE Trans Neural Netw May 2004;15(3):653–62.

[34] Kaur Livjeet, Joshi Mohinder Pal. Analysis of chemotaxis in bacterial foraging optimization algorithm. Int J Comput Appl May 2012;46(4):18–23. Full text available.

CHAPTER EIGHT

Final Remarks

Nowadays, bio-inspired algorithms have become a very important tool to solve engineering problems. Given that using classic optimization algorithms often implies computing and iteratively evaluating Hessians and gradients, the computational cost of this kind of algorithms—in some cases—may be excessively high. On the other hand, calculations of bio-inspired algorithms involve simple operations, most of them being linear equations. Furthermore, bio-inspired algorithms are easy to parallelize using a "divide and conquer" strategy, in which each individual in the population of the algorithm can be seen as a subprocess of the whole bio-inspired system. That means, each individual can be treated as a thread to solve a part of the problem because each individual of the population can be evaluated independently. If required, a phase of communication between individuals could be performed. So, using bio-inspired algorithms the computation complexity and execution time to solve engineering problems are often reduced, making them an efficient alternative to solve these kinds of problems.

It is important to mention that bio-inspired algorithms have a very explicit and easy-to-understand operation because they follow a simplified model of a biological entity. This makes their programming produce explicit codes and versions; handling change is also made simple.

The purpose of this chapter is to present a book summary with final remarks and general conclusions, and to offer a forecast regarding future work.

Several state-of-the-art applications to solve engineering problems, ranging from intelligent pattern recognition, object reconstruction, robot control and vision, intelligent identification, and control of nonlinear systems have been presented in this book. All of the applications presented are characterized by explicit, efficient, and robust systems, which involve using bio-inspired algorithms to function. As a result, the authors believe that using and designing bio-inspired algorithms is, and will, continue to be a strong and growing research area in engineering, and they hope that this book provides young researchers in this field inspiration to continue producing engineering solutions.

Chapter 1 offered a brief review of bio-inspiring algorithms and their importance to solving complex optimization problems in engineering.

© 2018 Elsevier Inc.
All rights reserved.

In addition, Chapter 1 briefly introduced bio-inspired algorithms used throughout the rest of the book. These included PSO, ABC, μABC, DE, and BFO.

Chapter 2 presented an approach to perform large-scale data classification using SVMs trained using KA combined with ABC, μABC, DE, and PSO. The combination of KA and bio-inspired algorithms allowed us to obtain a parallelized system of classification that was effectively applied to solve pattern recognition with data of large dimension, keeping its computational complexity lower than previous large-scale classifiers that use SVM. Our proposed SVM-bio-inspired algorithm can be used to solve problems as classification of chromosomes, spam filtering, information security, and other problems where the dimension and/or amount of data is very large and where the generalization of knowledge is highly desired to obtain a low training error.

Chapter 3 showed the design and implementation of a method to reconstruct 3D surfaces from point-clouds using RBF neural networks, which have been adjusted using PSO. Meshing functions in order to interpolate point-clouds to obtain compact surface representations is very important for CAD, robot mapping, and object description and recognition among other important applications. The results obtained using our algorithm show that although the obtained surfaces were not continuous, they can be used as compact descriptors for pattern recognition process and environmental mapping. This is because our proposal is fast enough to be implemented in real time, and the reduction of the number of parameters used to describe a shape with 3D point clouds is significant.

Chapter 4 presented an application of soft computing algorithms, such as PSO to solve important computer vision problems like image tracking and plane detection. These problem-solving capacities are very important since they can be used to perform more complex tasks, such as grasping and robot navigation for detection of obstacles and mapping. The presented image tracking algorithm is capable of working with large images. This is as a result of our having used an image pyramid search and PSO for looking for the best template match that improves the speed of the template search. The plane detection algorithm also used PSO with data obtained from an RGB-D sensor and used it to conduct a search in the image plane. However, complementing the data on the plane with depth information to construct planes allowed us to implement a robust algorithm for noise and outliers.

Chapter 5 dealt with a soft computing approach that is able to avoid obstacles and to move a robot to reach a goal. The approach is based on the PSO algorithm, where each particle represents a potential solution of a new position for the robot. Once the best particle of the actual iteration is selected, the robot is moved to the position that the best particle represents. This algorithm was tested with nonholonomic and holonomic robots, and it proved that bio-inspired algorithms are able to solve local navigation problems in real-time.

Chapter 6 presented the application of bio-inspired algorithms to improve neural identifiers for discrete-time unknown nonlinear systems. PSO was used to improve two kinds of neural identifiers. First, PSO was used to identify conditions conducive for an EKF learning algorithm to train a RHONN. The purpose of this undertaking was to identify a dynamic mathematical model to serve as a linear induction motor benchmark; second, the same enhanced PSO-EKF was used to train a recurrent multilayer perceptron in order to obtain an accurate neural model for forecasting in smart grids. Importance of these applications is attributable to the need of accurate dynamic models for modern purposes like control, forecasting, simulation, emulation among others. Besides these two applications shown, the proposed schemes are applicable to solve very different kinds of unknown nonlinear systems with noises, uncertainties, delays, saturations, etc.

Then, Chapter 7 presented the use of bio-inspired algorithms to improve neural controllers for discrete-time unknown nonlinear systems based on two approaches. First, a Neural-PSO Second-Order Sliding Mode Controller approach was presented to control a class of unknown nonlinear systems. Second, a Neural-BFO Second-Order Sliding Mode Controller for the same class of unknown nonlinear systems was incorporated. In order to show applicability of these controllers, they were applied to a Van der Pol Oscillator, and a comparative analysis was done to establish conclusions about the development of both controllers with respect to a traditional one. It was concluded that the Neural-BFO Second-Order Sliding Mode Controller had a better performance in trajectory tracking of a class of unknown discrete-time nonlinear systems with disturbances (external an internal).

The results of all the simulation and experimental work done with bio-inspired algorithms considered in this book demonstrate that through the use of these problem-solving approaches and tools adequate solutions may be obtained for a wide range of engineering problems; they also stand as proof of the importance and capabilities of the proposed methodologies.

Finally, we emphasize the highlights of the book:

- The proposed applications are very general in the sense that they are able to handle a large class of engineering problems.

- Applications considered in this book include both those of simulation and experimentation.
- Applications considered in this book are in the category of state-of-the-art, including intelligent pattern recognition, object reconstruction, robot control and vision, and intelligent identification and control of nonlinear systems, all of which use explicit, efficient and robust systems which involve using bio-inspired algorithms to function.

In regards to future work, it is possible to undertake the following:

- integration of these methodologies for autonomous robotic navigation,
- extrapolation of the proposed methodologies for cyberphysical systems,
- expansion of the proposed methodologies to complex systems,
- hybridization of bio-inspired algorithms with other soft computing techniques,
- improvement of well-known bio-inspired algorithms to avoid common problems like slow convergence and stagnation in local minimal among others, and
- development of new bio-inspired algorithms considering new developments for biological collective intelligence.

Finally, we are certain that this book will aid to rise above limitations linked to implementing bio-inspired algorithms to solve real-life challenges and complex engineering problems that are hard to be solved using traditional methodologies. Through the new generation of students and scientists applying the ideas proposed in this book both theoretically and practically, including in their research activities, the vital importance of all the concepts contained therein will be made evident.

INDEX

Printed in the United States
By Bookmasters